IEE CONTROL ENGINEERING SERIES 55

Series Editors: Professor D. P. Atherton
                Professor G. I. Irwin

Genetic algorithms in engineering systems

Other volumes in print in this series:

Genetic algorithms in engineering systems

Edited by
A. M. S. Zalzala
and
P. J. Fleming

The Institution of Electrical Engineers

Published by: The Institution of Electrical Engineers, London,
United Kingdom

© 1997: The Institution of Electrical Engineers

The Institution of Electrical Engineers,
Michael Faraday House,
Six Hills Way, Stevenage,
Herts. SG1 2AY, United Kingdom

British Library Cataloguing in Publication Data

A CIP catalogue record for this book
is available from the British Library

ISBN 0 85296 902 3

Printed in England by Bookcraft, Bath

Contents

Preface

Genetic algorithms (GAs) are general-purpose search and optimisation methods applicable to a wide variety of real-life problems. These algorithms are based on certain concepts of biological evolution, and represent the search space of a coded population of potential solutions. The population is then manipulated according to the survival of the fittest principle, hence converging to an optimal solution.

GAs can be applied in many fields but the theoretical innovations and practical applications in engineering systems are the subject of this volume. This book arises from the highly successful 1st IEE/IEEE international conference on *Genetic algorithms in engineering systems: innovations and applications* (GALESIA '95), held at the University of Sheffield, UK. Being the only one of its kind, the conference reflected the wide interest in the area, presenting new theories in optimisation, scheduling, control, system identification, neural networks and other fields. The applications side of the event included areas such as aerospace, robotics, utilities, signal processing and manufacturing.

The contributions presented in this book are extended versions of commissioned papers from some of the highest quality contributions to the conference. Chosen for their experience in the field, the authors are drawn from academia and industry worldwide. The chapters cover the main fields of work as well as presenting tutorial material in this important subject, which is currently receiving considerable attention from engineers.

The book starts with a broad survey of the current trends and techniques used in GAs, and the many variations from the original GA are discussed to illustrate how this powerful and versatile search and optimisation method is applicable to a broad range of activities. Chapter 1 also includes a brief discussion of the biological origins of GAs. Chapter 2 looks at the range of representation levels at which algorithms can be applied to intelligent control systems, including evolving control parameters, complex structures and rules. The focus of the chapter, however, is on evolving control rules embodied in the

SAMUEL system, a machine learning program that uses a GA and other competition-based heuristics to improve a set of control rules. Chapter 3 aims to illustrate how an existing GA can be modified and set up to explore the relevant trade-offs between multiple objectives with a minimum of effort. Most engineering problems are characterised by several noncommensurable and often competing objectives to be optimised, but usually have no unique, perfect solution due to the trade-offs involved. Simple experimental results are included for the purpose of illustration. Chapter 4 considers the problem of including multiple constraints in a large, ill-behaved search space, which is a common difficulty in all optimisation techniques applied in different areas, such as design, scheduling, system identification and control, or any of the myriad of areas to which genetic algorithms have been applied. In this chapter, this is addressed through a fuzzy logic method, reported as part of a genetic algorithm search.

Evolving the learning behaviours of neural networks is the topic covered by Chapter 5. It reports that the successful evolution of scaleable neural architectures is dependent on the encoding scheme, i.e. how the neural network is represented in the chromosome, and describes a simulation model that is capable of supporting learning behaviours in a unified fashion. Chapter 6 presents a new method to identify the parameters of nonlinear circuits, based on the concepts of synchronisation. This method is formulated as a global optimisation problem using a genetic algorithm. Three experimental examples are reported for estimating the five dimensionless parameters of the chaotic Chua's oscillator. The productivity of manufacturing processes is discussed in Chapter 7, with algorithms given for job shop scheduling. This is a much researched problem in operations research, where the objective is the optimal allocation of shared resources over time to competing activities. The authors illustrate their work with several benchmark problems.

Chapter 8 addresses the principles of the use of evolutionary algorithms in the motion planning of robotic systems. In addition, the implementation of these principles is reported for single mani-pulators, multiple arms and mobile arms. The authors emphasise the need for a tailor-fit design of the genetic structure (coding, parameters and objectives), and present comparative results for single and multiple objective optimisation formulations. Chapter 9 discusses the characteristics of aerodynamics through wing shape design problems, where it is demonstrated that distribution of the objective function can be extremely rough even in a simplified problem. Comparisons are reported with gradient-based optimisation, simulated annealing and GAs, as applied to the airfoil shape design using the approximation concept; the GA being the best for aerodynamic optimisation. Finally, Chapter 10 presents a genetic algorithm for a combinatorial optimisation problem: the design of VLSI macro cell layouts; the main feature being the total integration of the global routing into the placement process. Together with a hybrid method for the creation of the initial individuals, the approach contains detailed features with the chosen genotype representation being a tree with additional information for all nodes defining details for routing and sizing of the modules.

Ali Zalzala
Peter Fleming
(Sheffield 1997)

Contributors

Chapter 1
A Chipperfield
Department of Automatic Control
    and Systems Engineering
The University of Sheffield
Mappin Street
Sheffield
S1 3JD   UK

Chapter 2
J J Grefenstette
Code 5510
Naval Research Laboratory
Washington DC 20375   USA

Chapter 3
C M Fonseca and P J Fleming
Department of Automatic Control
    and Systems Engineering
The University of Sheffield
Mappin Street
Sheffield
S1 3JD   UK

Chapter 4
R Pearce
Rolls-Royce Applied Science
    Laboratory
PO Box 31
Derby
DE24 8BJ   UK

Chapter 5
S Lucas
Department of Electronic
    Systems Engineering
University of Essex
Wivenhoe Park
Colchester
CO4 3SQ   UK

Chapter 6
R Caponetto and M Lavorgna
SGS Thomson Microelectronics
Stradale Primosole 50
95121 Catania   Italy

L Fortuna and G Manganaro
DEES University of Catania
V.le A. Doria 6
95125 Catania   Italy

Chapter 7
T Yamada and R Nakano
NTT Communication Science
    Laboratories
2 Hikaridai, Seika-cho
Soraku-gun
Kyoto 619-02   Japan

Chapter 8
**A M S Zalzala, M C Ang, M Chen,
    A S Rana and G Wang**
Department of Automatic Control
    and Systems Engineering
The University of Sheffield
Mappin Street
Sheffield
S1 3JD   UK

Chapter 9
S Obayashi
Department of Aeronautics
    and Space Engineering
Tohoku University
Sendai 980-77   Japan

Chapter 10
V Schnecke
Department of Mathematics
    and Computer Science
University of Osnabruck
Albrechtstrasse 28
D-49069 Osnabruck   Germany

Chapter 1
Introduction to genetic algorithms

A. Chipperfield

In 1859 Charles Darwin (1809–82) published an extremely controversial book whose full title is *On the origin of species by means of natural selection, or the preservation of favoured races in the struggle for life,* which is now popularly known as *The origin of species.* He suggested that a species is continually developing, his controversial thesis implying that man himself came from ape-like stock. During his explorations, Darwin was impressed by the variations between species. He noticed that in almost all organisms there is a huge potential for the production of offspring as, for example, eggs and spores, but that only a small percentage survive to adulthood. He also observed that within a population there is a great deal of variation. This led him to deduce that those variants which survived the struggle to adulthood were, presumably, the ones most fit to do so. Supposing that individual variation could be inherited by offspring, Darwin saw evolution as the natural selection of inheritable variations.

Around the same time, Gregor Mendel (1822–84) investigated the inheritance of characteristics, or traits, in his experiments with pea plants. By examining hybrids from different strains of plant he obtained some notion of the interactions of characters. For example, when crossing tall plants with short ones, all the resulting hybrids were tall regardless of which plant donated the pollen. Mendel declared that the character, or gene as they later came to be known, for the tall plant was dominant and that the gene for shortness was recessive. Although Mendel's experiments laid the foundations for the study of genetics, it was not until 30 years after his death that Walter Sutton (1877–1916) discovered that genes were part of chromosomes in the nucleus.

However, Darwin's theory emphasised the role of continuous variation within species. In contrast, distinct differences between species are not uncommon in nature, i.e. discontinuous variation. Hugo de Varis (1848–1935) observed that in a population of cultivated plants, strikingly different variants would occasionally appear. To explain this discontinuous variation, de Varis developed a theory of mutation. Superficially, the new science of genetics seemed to support the mutation theory of evolution against orthodox Darwinism. With greater understanding of the structure of genes, geneticists came to realise how subtle the effects of mutation could be. If a characteristic is determined by a single gene, mutation may have a dramatic effect; but if a battery of genes combines to control that characteristic, mutation in one of them may only have a negligible effect. It is clear, therefore, that there is not a sharp distinction between mutation and Darwinian theory of evolution as they overlap. The principle of selection does, however, remain sound.

Genetic algorithms (GAs) are stochastic search and optimisation methods based on the metaphors of natural biological evolution described above. Broadly, they are part of the larger class of evolutionary algorithms (EAs) [1] which also includes evolutionary programming (EP) [2], evolution strategies (ES) [3] and genetic programming (GP) [4]. EAs operate with a population of potential solutions to a problem, applying the principles of survival of the fittest, reproduction and mutation to producing successively better approximations to the solution. At each iteration of an EA, a new generation of approximations is created by the processes of selection and reproduction leading to the evolution of populations of individuals which are better suited to their environment – the problem domain – than the individuals from which they were created, just as it occurs in natural adaptation.

This Chapter starts with an overview of the basic mechanics of GAs and highlights their major differences when compared to traditional and enumerative search and optimisation techniques. The main components of the GA are then described in some detail and various alternative approaches to the major procedures are considered. After a brief discussion of other evolutionary algorithms, parallel models of the GA are then considered and it is shown how it may be possible to improve the performance of the algorithm even when it is implemented on a sequential computer. Next, considerations commonly arising in engineering systems and the manner in which they may be treated through the application of GAs are discussed. Finally, an example of the use of a GA in aircraft engine controller configuration design is presented to demonstrate how GAs may be applied to problems for which there are currently no other direct methods of solution.

1.1 What are genetic algorithms?

The GA is a stochastic global search method that mimics the metaphor of natural biological evolution [5]. GAs operate on a population of potential solutions applying the principle of survival of the fittest to produce (hopefully) better and better approximations to a solution. At each generation, a new set of approximations is created by the process of selecting individuals according to their level of fitness in the problem domain and breeding them together using operators borrowed from natural genetics. This process leads to the evolution of populations of individuals which are better suited to their environment than the individuals that they were created from.

1.1.1 Overview of GAs

Individuals, or current approximations, are encoded as strings, chromosomes, composed over some alphabet(s), so that the genotypes (chromosome values) are uniquely mapped onto the decision variable (phenotypic) domain. The most commonly used representation in GAs is the binary alphabet {0, 1} although other representations can be used, e.g. ternary, integer, real-valued etc. For example, a problem with two variables, x_1 and x_2, may be mapped onto the chromosome structure in the following way:

$$1\ 0\ 0\ 1\ 0\ 1\ 1\ 0\ 1\ 1\ \vdots\ 0\ 1\ 0\ 0\ 1\ 0\ 1\ 1\ 1\ 0\ 1\ 1\ 0\ 1$$

$$\underset{\xleftarrow{\hspace{3cm}}}{x_1} \qquad \underset{\xleftarrow{\hspace{5cm}}}{x_2}$$

where x_1 is encoded with ten bits and x_2 with 15 bits, possibly reflecting the level of accuracy or range of the individual decision variables. Examining the chromosome string in isolation yields no information about the problem which we are trying to solve. It is only with the decoding of the chromosome into its phenotypic values that any meaning can be applied to the representation. However, as described below, the search process will operate on this encoding of the decision variables, rather than the decision variables themselves, except, of course, where real-valued genes are used.

Having decoded the chromosome representation into the decision variable domain, it is possible to assess the performance, or fitness, of individual members of a population. This is done through an objective function that characterises an individual's performance in the problem domain. In the natural world, this would be an individual's ability to survive in its present environment. Thus, the objective function establishes the basis for selection of pairs of individuals which will be mated together during reproduction.

During the reproduction phase, each individual is assigned a fitness value derived from its raw performance measure given by the objective function. This value is used in the selection process to bias it towards fitter individuals. Highly fit individuals, relative to the whole population, have a high probability of being selected for mating whereas less fit individuals have a correspondingly low probability of being selected.

Once the individuals have been assigned a fitness value, they can be chosen from the population, with a probability according to their relative fitness, and recombined to produce the next generation. Genetic operators manipulate the characters (genes) of the chromosomes directly, using the assumption that certain individual's gene codes, on average, produce fitter individuals. The recombination operator is used to exchange genetic information between pairs, or larger groups, of individuals. The simplest recombination operator is that of single-point crossover.

Consider the two parent binary strings:

$$P_1 = 0\ 0\ 0\ 1\ 1\ 1\ 1\ 0$$
$$P_2 = 1\ 0\ 1\ 0\ 0\ 0\ 1\ 1$$

If an integer position, i, is selected uniformly at random from the range $[1, l\text{-}1]$, where l is the string length, and the genetic information exchanged between the individuals about this point, then two new offspring strings are produced. The two offspring below are produced when the crossover point $i = 4$ is selected:

$$O_1 = 0\ 0\ 0\ 1\ 0\ 0\ 1\ 1$$
$$O_2 = 1\ 0\ 1\ 0\ 1\ 1\ 1\ 0$$

This crossover operation is not necessarily performed on all strings in the population. Instead, it is applied with a probability P_x when the pairs are chosen for breeding. A further genetic operator, called mutation, is then applied to the new chromosomes, again with a set probability, P_m. Mutation causes the individual genetic representation to be changed according to some probabilistic rule. In the binary string representation, mutation will cause a random bit to change its state, $0 \Rightarrow 1$ or $1 \Rightarrow 0$. So, for example, mutating the seventh bit of O_1 leads to the new string:

$O_{1m} = 0\ 0\ 0\ 1\ 0\ 0\ 0\ 1$

After recombination and mutation, the individual strings are then, if necessary, decoded, the objective function evaluated, a fitness value assigned to each individual and individuals selected for mating according to their fitness, and so the process continues through subsequent generations. In this way the average performance of individuals in a population is expected to increase, as good individuals are preserved and bred with one another and the less fit individuals die out. The GA is terminated when some criteria are satisfied, e.g. a certain number of generations completed, a mean deviation in the performance of individuals in the population or when a particular point in the search space is encountered.

1.1.2 GAs versus traditional methods

From the above discussion, it can be seen that the GA differs substantially from more traditional search and optimisation methods. The four most significant differences are:

- GAs search a population of points in parallel, not a single point.
- GAs use probabilistic transition rules, not deterministic ones.
- GAs work on an encoding of the parameter set rather than the parameter set itself (except where real-valued individuals are used).
- GAs do not require derivative information or other auxiliary knowledge; only the objective function and corresponding fitness levels influence the directions of search.

It is important to note that the GA can provide a number of potential solutions to a given problem and the choice of final solution is left to the user. In cases where a particular problem does not have a unique solution, for example in multiobjective optimisation where the result is usually a family of Pareto-optimal solutions, the GA is potentially useful for identifying these alternative solutions simultaneously.

1.2 Major elements of the GA

The simple genetic algorithm (SGA) is described by Goldberg [6] and is used here to illustrate the basic components of the GA. A pseudocode outline of the SGA is shown in Figure 1.1. The population at time t is represented by the time-dependent variable *P*, with the initial population of random estimates being *P*(0). Using this outline of a GA, the remainder of this Section describes the major elements of the GA.

```
procedure GA
begin
        t = 0;
        initialize P(t);
        evaluate P(t);
        while not finished do
        begin
                t = t + 1;
                select P(t) from P(t-1);
                reproduce pairs in P(t);
                evaluate P(t);
        end
end.
```

Figure 1.1   A simple genetic algorithm

1.2.1 Population representation and initialisation

GAs operate simultaneously on a number of potential solutions, called a population, consisting of some encoding of the parameter set. Typically, a population is composed of between 30 and 100 individuals, although, a variant called the micro GA uses very small populations, ~ten individuals, with a restrictive reproduction and replacement strategy in an attempt to satisfy real-time execution requirements [7].

The most commonly used representation of chromosomes in the GA is that of the single-level binary string. Here, each decision variable in the parameter set is encoded as a binary string and these are concatenated to form a chromosome (see the example in Section 1.1). The use of Gray coding has been advocated as a method of overcoming the hidden representational bias in conventional binary representation as the Hamming distance between adjacent values is constant [8]. Empirical evidence of Caruana and Schaffer [9] suggests that large Hamming distances in the representational mapping between adjacent values, as is the case in the standard binary representation, can result in the search process being deceived or unable to efficiently locate the global minimum. A further approach of Schmitendorgf *et al.* [10], is the use of logarithmic scaling in the conversion of binary-coded chromosomes to their real phenotypic values. Although the precision of the parameter values is possibly less consistent over the desired range, in problems where the spread of feasible parameters is unknown a larger search space may be covered with the same number of bits than when using a linear mapping scheme, thus allowing the computational burden of exploring unknown search spaces to be reduced to a more manageable level.

Although binary-coded GAs are most commonly used, there is an increasing interest in alternative encoding strategies, such as integer and real-valued representations. For some problem domains, it is argued that the binary representation is in fact deceptive in that it obscures the nature of the search [11]. In the subset selection problem [12], for example, the use of an integer representation and look-up tables provides a convenient and natural way of expressing the mapping from representation to problem domain. Consider the travelling salesperson problem, the task being to find the shortest route visiting all the cities from a given set exactly once. By using integer labels, each candidate solution can be uniquely represented as a permutation of these elements. For example, in a seven-city tour, both {2, 7, 1, 3, 5, 6, 4} and {6, 4, 7, 1, 5, 3, 2} represent paths between the cities. Thus, the chromosomes used in a GA to solve this problem would contain seven integers, each integer corresponding to a city in the tour.

The use of real-valued genes in GAs is claimed by Wright [13] to offer a number of advantages in numerical function optimisation over binary encodings. Efficiency of the GA is increased as there is no need to convert chromosomes to phenotypes before each function evaluation, less memory is required as efficient floating point internal computer representations can be used directly, there is no loss in precision by discretisation to binary or other values and there is greater freedom to use different genetic operators. The use of real-valued encodings is described in detail by Michalewicz [14] and in the literature on evolution strategies (see, for example, [15]).

Having decided on the representation, the first step in the SGA is to create an initial population. This is usually achieved by generating the required number of individuals using a random number generator which uniformly distributes numbers in the desired range. For example, with a binary population of N_{ind} individuals whose chromosomes are L_{ind} bits long, $N_{ind} \times L_{ind}$ random numbers uniformly distributed from the set {0, 1} would be produced.

A variation is the extended random initialisation procedure of Bramlette [11] whereby a number of random initialisations are tried for each individual and the one with the best performance is chosen for the initial population. Other users of GAs have seeded the initial population with some individuals that are known to be in the vicinity of the global minimum (see, for example, [16] and [17]). This approach is, of course, only applicable if the nature of the problem is well understood beforehand or if the GA is used in conjunction with a knowledge based system.

1.2.2 The objective and fitness functions

The objective function is used to provide a measure of how individuals have performed in the problem domain. In the case of a minimisation problem, the most fit individuals will have the lowest numerical value of the associated objective function. This raw measure of fitness is usually only used as an intermediate stage in determining the relative performance of individuals in a GA. Another function, the fitness function, is normally used to transform the objective function value into a measure of relative fitness [18], thus:

$$F(x)=g(f(x))$$

where f is the objective function, g transforms the value of the objective function to a non-negative number and F is the resulting relative fitness. This mapping is always necessary when the objective function is to be minimised as the lower objective function values correspond to fitter individuals. In many cases, the fitness function value corresponds to the number of offspring which an individual can expect to produce in the next generation. A commonly used transformation is that of proportional fitness assignment (see, for example, [6]). The individual fitness, $F(x_i)$, of each individual is computed as the individual's raw performance, $f(x_i)$, relative to the whole population, i.e.:

$$F(x_i)=\frac{f(x_i)}{\sum_{i=1}^{Nind}f(x_i)}$$

where N_{ind} is the population size and x_i is the phenotypic value of individual i. Although this fitness assignment ensures that each individual has a probability of reproducing according to its relative fitness, it fails to account for negative objective function values.

A linear transformation which offsets the objective function [6] is often used prior to fitness assignment, such that:

$$F(x) = af(x) + b$$

where *a* is a positive scaling factor if the optimisation is maximising and negative if we are minimising. The offset *b* is used to ensure that the resulting fitness values are non-negative.

The use of linear scaling and offsetting outlined above is, however, a possible cause of rapid convergence. The selection algorithm (see below) selects individuals for reproduction on the basis of their relative fitness. Using linear scaling, the expected number of offspring is approximately proportional to that individual's performance. As there is no constraint on an individual's performance in a given generation, highly fit individuals in early generations can dominate the reproduction causing rapid convergence to possibly suboptimal solutions. Similarly, if there is little deviation in the population, then scaling provides only a small bias towards the most fit individuals.

A further method of transforming the objective function values to fitness measures is power law scaling [6]. Here, the scaled fitness is taken as some specified power, *k*, of the raw fitness, *f*:

$$F(x) = f(x)^k$$

The value of *k* is, in general, problem dependent and may be dynamically changed during the execution of the GA to shrink or stretch the range of fitness measures as required.

Baker [19] suggests that limiting the reproductive range, so that no individuals generate an excessive number of offspring, prevents premature convergence. Here, individuals are assigned a fitness according to their rank in the population rather than their raw performance. One variable, *SP*, is used to determine the bias, or selective pressure, towards the most fit individual and the fitness of the others is determined by:

$$F(x_i)=2- SP+2(SP-1)\ \frac{x_i - 1}{N_{ind}-1}$$

where x_i is the position in the ordered population of individual *i*.

For example, for a population size of $N_{ind} = 40$ and selective pressure of *SP* = 1.1, individuals are given a fitness value in the range [0.9, 1.1]. The least fit individual has a fitness of 0.9 and the most fit is assigned a fitness of 1.1. The increment in the fitness value between adjacent ranks is thus 0.0051.

1.2.3 Selection

Selection is the process of determining the number of times, or trials, a particular individual is chosen for reproduction and, thus, the number of offspring that an individual will produce. The selection of individuals can be viewed as two separate processes:

(1) determination of the number of trials an individual can expect to receive;
(2) conversion of the expected number of trials into a discrete number of offspring.

The first part is concerned with the transformation of raw fitness values into a real-valued expectation of an individual's probability of reproducing and is dealt with in the previous subsection as fitness assignment. The second part is the probabilistic selection of individuals for reproduction based on the fitness of individuals relative to one another, and is sometimes known as sampling. The remainder of this subsection will review some of the more popular selection methods in current usage.

Baker [20] presented three measures of performance for selection algorithms, bias, spread and efficiency. Bias is defined as the absolute difference between an individual's actual and expected selection probability. Optimal zero bias is therefore achieved when an individual's selection probability equals its expected number of trials. Spread is the range in the possible number of trials that an individual may achieve. If $f(i)$ is the actual number of trials that individual i receives, then the minimum spread is the smallest spread that theoretically permits zero bias, i.e:

$$f(i) \in \{\lfloor et(i) \rfloor, \lceil et(i) \rceil\}$$

where $et(i)$ is the expected number of trials of individual i, $\lfloor et(i) \rfloor$ is the floor of $et(i)$ and $\lceil et(i) \rceil$ is the ceiling. Thus, although bias is an indication of accuracy, the spread of a selection method measures its consistency.

The desire for efficient selection methods is motivated by the need to maintain a GA's overall time complexity. It has been shown in the literature that the other phases of a GA (excluding the actual objective function evaluations) are $O(L_{ind}.N_{ind})$ or better time complexity. The selection algorithm should thus achieve zero bias while maintaining a minimum spread and not contributing to an increased time complexity of the GA.

1.2.3.1 Roulette wheel selection methods

Many selection techniques employ a roulette wheel mechanism to probabilistically select individuals based on some measure of their performance. A real-valued interval, *Sum*, is determined as either the sum of the individuals' expected selection probabilities or the sum of the raw fitness values over all the individuals in the current population. Individuals are then mapped one-to-one into contiguous intervals in the range [0, *Sum*]. The size of each individual interval corresponds to the fitness value of the associated individual. For example, in Figure 1.2 the circumference of the roulette wheel is the sum of all six individuals' fitness values. Individual 5 is the most fit individual and occupies the largest interval, whereas individuals 6 and 4 are the least fit and have correspondingly smaller intervals within the roulette wheel. To select an individual, a random number is generated in the interval [0, *Sum*] and the individual whose segment spans the random number is selected. This process is repeated until the desired number of individuals have been selected.

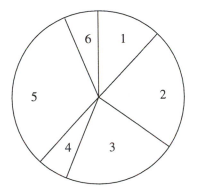

Figure 1.2   Roulette wheel selection

The basic roulette wheel selection method is stochastic sampling with replacement (SSR). Here, the segment size and selection probability remain the same throughout the selection phase and individuals are selected according to the procedure outlined above. SSR gives zero bias but a potentially unlimited spread. Any individual with a segment size > 0 could entirely fill the next population.

Stochastic sampling with partial replacement (SSPR) extends upon SSR by resizing an individual's segment if it is selected. Each time an individual is selected, the size of its segment is reduced by 1·0. If the segment size becomes negative, then it is set to 0·0. This provides an upper bound on the spread of $\lceil et(i) \rceil$. However, the lower bound is zero and the bias is higher than that of SSR.

Remainder sampling methods involve two distinct phases. In the integral phase, individuals are selected deterministically according to the integer part of their expected trials. The remaining individuals are then selected probabilistically from the fractional part of the individual's expected values. Remainder stochastic sampling with replacement (RSSR) uses roulette wheel selection to sample the individual not assigned deterministically. During the roulette wheel selection phase, individuals' fractional parts remain unchanged and, thus, compete for selection between spins. RSSR provides zero bias and the spread is lower bounded. The upper bound is limited only by the number of fractionally assigned samples and the size of the integral part of an individual. For example, any individual with a fractional part > 0 could win all the samples during the fractional phase. Remainder stochastic sampling without replacement (RSSWR) sets the fractional part of an individual's expected values to zero if it is sampled during the fractional phase. This gives RSSWR minimum spread, although this selection method is biased in favour of smaller fractions.

1.2.3.2 Stochastic universal sampling

Stochastic universal sampling (SUS) is a single-phase sampling algorithm with minimum spread and zero bias. Instead of the single selection pointer employed in roulette wheel methods, SUS uses N equally spaced pointers, where N is the number of selections required. The population is shuffled randomly and a single random number in the range $[0, Sum/N]$ is generated, ptr. The N individuals are then chosen by generating the N pointers spaced by 1, $[ptr, ptr + 1, ..., ptr + N–1]$, and selecting the individuals whose fitnesses span the positions of the pointers. An individual is thus guaranteed to be selected a minimum of times $\lfloor et(i) \rfloor$ and no more than $\lceil et(i) \rceil$, thus achieving minimum spread. In addition, as individuals are selected entirely on their position in the population, SUS has zero bias. For these reasons, SUS has become one of the most widely used selection algorithms in current GAs.

1.2.4 Crossover (recombination)

The basic operator for producing new chromosomes in the GA is that of crossover. Like its counterpart in nature, crossover produces new individuals which have some parts of both parents' genetic material. The simplest form of crossover is that of single-point crossover, described in the overview of GAs in Section 1.2.1. In this Section, a number of variations on crossover are described and discussed and the relative merits of each reviewed.

1.2.4.1 Multipoint crossover

For multipoint crossover, m crossover positions, $k_i \in \{1, 2, ..., l-1\}$, where k_i are the crossover points and l is the length of the chromosome, are chosen at random with no duplicates and sorted into ascending order. Then, the bits between successive crossover points are exchanged between the two parents to produce two new offspring. The section between the first allele position and the first crossover point is not exchanged between individuals. This process is illustrated in Figure 1.3.

The idea behind multipoint, and indeed many of the variations on

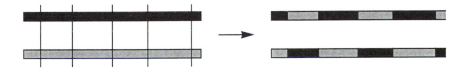

Figure 1.3   Multipoint crossover

the crossover operator, is that the parts of the chromosome representation that contribute most to the performance of a particular individual may not necessarily be contained in adjacent substrings [21]. Further, the disruptive nature of multipoint crossover appears to encourage the exploration of the search space, rather than favouring convergence to highly fit individuals early in the search, thus making the search more robust [22].

1.2.4.2 Uniform crossover

Single and multipoint crossover define cross points as places between loci where a chromosome can be split. Uniform crossover [23] generalises this scheme to make every locus a potential crossover point. A crossover mask, the same length as the chromosome structures, is created at random and the parity of the bits in the mask indicates which parent will supply the offspring with which bits. Consider the following two parents, crossover mask and resulting offspring:

$$
\begin{aligned}
P_1 \quad &= 1\ 0\ 1\ 1\ 0\ 0\ 0\ 1\ 1\ 1 \\
P_2 \quad &= 0\ 0\ 0\ 1\ 1\ 1\ 1\ 0\ 0\ 0 \\
\text{Mask} \quad &= 0\ 0\ 1\ 1\ 0\ 0\ 1\ 1\ 0\ 0 \\
O_1 \quad &= 0\ 0\ 1\ 1\ 1\ 1\ 0\ 1\ 0\ 0 \\
O_2 \quad &= 1\ 0\ 0\ 1\ 0\ 0\ 1\ 0\ 1\ 1
\end{aligned}
$$

Here, the first offspring, O_1, is produced by taking the bit from P_1 if the corresponding mask bit is 1, or the bit from P_2 if the corresponding mask bit is 0. Offspring O_2 is created using the inverse of the mask or, equivalently, swapping P_1 and P_2.

Uniform crossover, like multipoint crossover, has been claimed to reduce the bias associated with the length of the binary representation used and the particular coding for a given parameter set. This helps to overcome the bias in single-point crossover towards short substrings without requiring precise understanding of the significance of individual bits in the chromosome representation. Spears and De Jong [24] have demonstrated how uniform crossover may be parameterised by applying a probability to the swapping of bits. This extra parameter can be used to control the amount of disruption during recombination without introducing a bias towards the length of the representation used. When uniform crossover is used with real-valued alleles, it is usually referred to as discrete recombination.

1.2.4.3 Other crossover operators

A related crossover operator is that of shuffle [25]. A single cross point is selected, but before the bits are exchanged, they are randomly shuffled in both parents. After recombination, the bits in the offspring are unshuffled. This too removes positional bias as the bits are randomly reassigned each time crossover is performed.

The reduced surrogate operator [21] constrains crossover to always produce new individuals wherever possible. Usually, this is implemented by restricting the location of crossover points such that crossover points only occur where gene values differ.

1.2.4.4 Intermediate recombination

Given a real-valued encoding of the chromosome structure, intermediate recombination is a method of producing new phenotypes around and between the values of the parents' phenotypes [26]. Offspring are produced according to the rule:

$$O_1 = P_1 \times \alpha(P_2 - P_1)$$

where α is a scaling factor chosen uniformly at random over some interval, typically [-0.25, 1.25] and P_1 and P_2 are the parent chromosomes (see, for example, [26]). Each variable in the offspring is the result of combining the variables in the parents according to the above expression with a new α chosen for each pair of parent genes. In geometric terms, intermediate recombination is capable of producing new variables within a slightly larger hypercube than that defined by the parents but constrained by the range of α as shown in Figure 1.4(a).

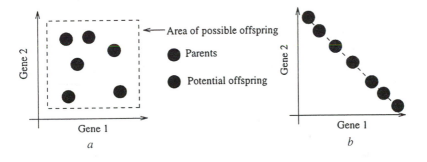

Figure 1.4   *Geometric effect of recombination operators*
  a intermediate recombination
  b line recombination

1.2.4.5 Line recombination

Line recombination [26] is similar to intermediate recombination, except that only one value of α is used in the recombination. Figure 1.4(b) shows how line recombination can generate any point on the line defined by the parents within the limits of the perturbation, α, for a recombination in two variables.

1.2.4.6 Discussion

The binary operators discussed in this Section have all, to some extent, used disruption in the representation to help improve exploration during recombination. Although these operators may be used with real-valued populations, the resulting changes in the genetic material after recombination would not extend to the actual values of the decision variables, although offspring may, of course, contain genes from either parent. The intermediate and line recombination operators overcome this limitation by acting on the decision variables themselves. Like uniform crossover, the real-valued operators may also be parameterised to provide a control over the level of disruption introduced into offspring.

For discrete-valued representations, variations on the recombination operators may be used that ensure that only valid values are produced as a result of crossover [27]. In particular, for problems such as bin packing and graph colouring, it has been shown that crossover operators which exploit domain specific knowledge may be employed to improve the efficiency of the GA [28]. The wider application of evolutionary computing methods to a greater range of problem domains has resulted in a large number of special purpose operators. However, as Baxter *et al.* report [29], repeated experimentation with operator probabilities may be necessary if any real benefit is to be gained from their use.

1.2.5 Mutation

In natural evolution, mutation is a random process where one allele of a gene is replaced by another to produce a new genetic structure. In GAs, mutation is randomly applied with low probability, typically in the range 0.001 and 0.01, and modifies elements in the chromosomes. Usually considered as a background operator, the role of mutation is often seen as providing a guarantee that the probability of searching any given string will never be zero and acting as a safety net to recover good genetic material which may be lost through the action of selection and crossover [6].

The effect of mutation on a binary string is illustrated in Figure 1.5 for a ten-bit chromosome representing a real value decoded over the interval $[0, 10]$ using both standard and Gray coding and a mutation point of three in the binary string. Here, binary mutation flips the value of the bit at the loci selected to be the mutation point. The effect of mutation on the decision variable, of course, depends on the encoding scheme used. Given that mutation is generally applied uniformly to an entire population of strings, it is possible that a given binary string may be mutated at more than one point.

mutation point											binary	Gray
Original string	0	0	0	1	1	0	0	0	1	0	0.9659	0.6634
Mutated string	0	0	0	1	1	0	0	0	1	0	2.2146	1.8439

Figure 1.5   Binary mutation

With nonbinary representations, mutation is achieved by either perturbing the gene values or random selection of new values within the allowed range. Wright [13] and Janikow and Michalewicz [30] demonstrate how real-coded GAs may take advantage of higher mutation rates than binary-coded GAs, increasing the level of possible exploration of the search space without adversely affecting the convergence characteristics. Indeed, Tate and Smith [31] argue that for codings more complex than binary, high mutation rates can be both desirable and necessary and show how, for a complex combinatorial optimisation problem, high mutation rates and non-binary coding yielded significantly better solutions than the normal approach.

Many variations on the mutation operator have been proposed. For example, biasing the mutation towards individuals with lower fitness values to increase the exploration in the search without losing information from the fitter individuals [32] or parameterising the mutation such that the mutation rate decreases with the population convergence [33]. Mühlenbein and Schlierkamp-Voosen [26] have introduced a mutation operator for the real-coded GA that uses a non-linear term for the distribution of the range of mutation applied to gene values. It is claimed that by biasing mutation towards smaller changes in gene values, mutation can be used in conjunction with recombination as a foreground search process. Other mutation operations include that of trade mutation [12], whereby the contribution of individual genes in a chromosome is used to direct mutation towards weaker terms, and reorder mutation [12], which swaps the positions of bits or genes to increase diversity in the decision variable space.

1.2.6 Reinsertion

Once a new population has been produced by selection and recom-bination of individuals from the old population, the fitness of the individuals in the new population may be determined. If fewer individuals are produced by recombination than the size of the original population, then the fractional difference between the new and old population sizes is termed a generation gap [34]. In the case where the number of new individuals produced at each generation is one or two, the GA is said to be steady state [35] or incremental [36]. If one or more of the most fit individuals is deterministically allowed to propagate through successive generations then the GA is said to use an elitist strategy.

To maintain the size of the original population, the new individuals have to be reinserted into the old population. Similarly, if not all the new individuals are to be used at each generation or if more offspring are generated than the size of the old population then a reinsertion scheme must be used to determine which individuals are to exist in the new population. An important feature of not creating more offspring than the current population size at each generation is that the generational computational time is reduced, most dramatically in the case of the steady-state GA, and that the memory requirements are smaller as fewer new individuals need to be stored while offspring are produced.

When selecting which members of the old population should be replaced the most apparent strategy is to replace the least fit members deterministically. However, in studies, Fogarty [37] has shown that no significant difference in convergence characteristics was found when the individuals selected for replacement were chosen with inverse proportional selection or deterministically as the least fit. He further asserts that replacing the least fit members effectively implements an elitist strategy as the most fit will probabilistically survive through successive generations. Indeed, the most successful replacement scheme was one that selected the oldest members of a population for replacement. This is reported as being more in keeping with generational reproduction as every member of the population will, at some time, be replaced. Thus, for an individual to survive successive generations, it must be sufficiently fit to ensure propagation into future generations.

1.2.7 Termination of the GA

Because the GA is a stochastic search method, it is difficult to formally specify convergence criteria. As the fitness of a population may remain static for a number of generations before a superior individual is found, the application of conventional termination criteria becomes problematic. A common practice is to terminate the GA after a prespecified number of generations and then test the quality of the best members of the population against the problem definition. If no acceptable solutions are found, the GA may be restarted or a fresh search initiated.

1.3 Other evolutionary algorithms

Although similar at the highest level, many variations exist in EAs. Evolutionary programming (EP) [2], arising from the desire to generate machine intelligence, typically uses a representation tailored to the problem domain. For example, in numerical optimisation vectors of real-valued numbers would be used and combinatorial problems would employ ordered lists. Given a population size of N, all N individuals are selected as parents and a representation specific mutation operator used to generate N offspring. The N offspring would then be evaluated and the next generation selected using a fitness-based probabilistic function on these $2N$ individuals. The mutation operator in EP is often adaptive and different adaptation rates may be used for each decision variable within an individual.

Evolution strategies (ES) [3] originally employed a population size of one individual, mutation and selection. Schwefel [38] used real-valued representation for individuals and introduced recombination and population sizes of greater than one individual. Parents are randomly selected and recombination and mutation used to produce more than N offspring. Selection is then used to select either the N best offspring or the N best individuals from the parents and offspring to make up the next generation. Unlike EP, recombination is an important operator in ES.

Genetic algorithms (GAs) traditionally use the more domain independent binary representation although other representations are now being employed. Selection of parents is probabilistic, based on a fitness function and N children are produced from the N parents using mutation and recombination operators. These offspring are then the new population. In GAs, recombination is considered to be the primary operator and mutation a background process. Genetic programming [4] uses EAs to evolve more complex structures such as Lisp expressions or neural networks to solve specific problems. For a comprehensive discussion of the differences between the various EAs, the interested reader is referred to [1].

1.4 Parallel GAs   *for multiobjective*

Given the preceding description of the GA, it is clear that the GA may be parallelised in a number of ways. Indeed, there are numerous variations on parallel GAs, many of which are very different from the original GA presented by Holland [5]. Most of the major differences are encountered in the population structure and the method of selecting individuals for reproduction. The motivation for exploring parallel GAs is manifold. One may wish to improve speed and efficiency by employing a parallel computer, apply the GA to larger problems or try to follow the biological metaphor more closely by introducing structure and geographic location into the population. As this Section will show, the benefits of using parallel GAs, even when run on a sequential machine, can be more than just a speed up in the execution time.

In deciding whether a parallel GA is useful for a specific problem, the trade off between population diversity, in terms of the population size, versus the execution time needs to be considered. For example, a small population size will yield a short execution time but may mean that some areas of the search space are not investigated because of the lack of genetic diversity in the population. This may mean that only suboptimal solutions are found, or, in the worst case, no satisfactory solutions are obtained. Large populations, on the other hand, may maintain diversity in the population but at the expense of execution time. Depending on the nature of the problem being addressed, excessive diversity may also mean that the GA is unable to find a satisfactory solution due to selective pressure driving reproduction towards suboptimal points. In many cases, due to the structure of the population and the use of local selection rules, parallel GAs offer an attractive mechanism for allowing diversity to exist within a population without unduly affecting the convergence characteristics of the GA. For some classes of problem, for example those characterised by multimodal search spaces or multiobjective formulations, the parallel GA can be shown to be more effective than the sequential GA, allowing multiple, equally satisfactory, solution estimates to coexist in the global population simultaneously.

In the remainder of this Section we describe a number of parallel GAs and use three broad categories to classify them: global, migration and diffusion. These categories reflect the different ways in which parallelism is exploited in the GA and the nature of the population structure and recombination mechanisms used. The global GA treats the entire population as a single breeding unit and aims to exploit the algorithmic parallelism inherent in the GA. Migration GAs divide the population into a number of subpopulations, each of which is treated as a separate breeding unit under the control of a conventional GA. To encourage the proliferation of good genetic material throughout the whole population, individuals migrate between the sub-populations from time to time. Generally, the migration GA is considered coarse grained. The diffusion GA treats each individual as a separate breeding unit, the individuals it may mate with being selected from within a small local neighbourhood. The use of local selection and reproduction rules leads to a continuous diffusion of individuals over the population. The diffusion GA is usually considered fine grained.

1.4.1 Global GAs

Examination of the pseudocode outline of the sequential simple GA given in Figure 1.1 reveals that a significant proportion of the computation in a GA is composed of taking pairs of individuals, combining them to form new offspring, applying mutation and evaluating a cost function. Taking a population size of, say, 50 and assuming that reproduction of two individuals creates two new offspring, then the inner loop of Figure 1.1 contains 25 discrete operations that may be performed concurrently at each generation. The worker/farmer architecture in Figure 1.6 demonstrates how this geometric parallelism may be exploited by a parallel computer.

The GA farmer node initialises and holds the entire population, performs selection and assigns fitness to individuals. The worker nodes recombine individuals, apply mutation and evaluate the objective function for the resulting offspring. Goldberg [6] describes a similar scheme, the synchronous master-slave, whereby a hybrid GA uses a local search routine at each worker to further refine the estimates generated at each node. Others, notably Fogarty and Huang [36] and Dodd *et al.* [39] use the processor farm for the evaluation of objective functions only.

Although the farmed GA does not embrace all of the parallelism inherent in the GA, near linear speed up has been reported in cases where the objective function is significantly more computationally expensive than the GA itself. In particular, when the objective function being minimised is of low computational cost, then there is potentially a bottleneck at the farmer while fitness assignment and selection are performed. The computational efficiency, of course, depends on the balance between the cost of the parallel parts of the GA and the sequential elements. Thus, the farmed GA may be inefficient if the objective function evaluation times vary greatly.

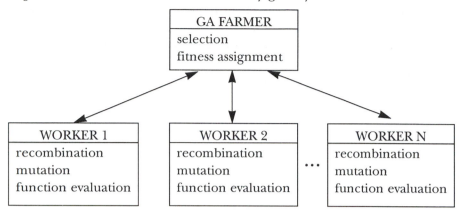

Figure 1.6   A worker/farmer GA

Goldberg [6] also describes a semisynchronous master-slave GA that overcomes this potential bottleneck by relaxing the requirement for strict synchronous operation. Here, individuals are selected and inserted into the population as and when the worker nodes complete their tasks. However, both the synchronous and semisynchronous models are potentially unreliable because of the dependence on the single farmer process.

A more robust extension to the worker/farmer implementations is the asynchronous, concurrent GA [6]. Using a number of identical processors, genetic operators and objective function evaluations are performed independently of one another on a population stored in a shared memory. This requires that no individual be accessed by more than one processor simultaneously. Although more complicated to implement than the conventional farmed GAs described above, this scheme is highly tolerant of processor and memory failure. Even if only one processor and some of the shared memory are functioning, it is still possible for useful processing to be performed.

1.4.2 Migration GAs

The GA as described thus far operates globally on a single population. That is, individuals are processed probabilistically on their performance in the population as a whole and any individual has the potential to mate with any other individual in the entire population.

This treatment of the population as a single breeding unit is known as panmixia.

In natural evolution, species tend to reproduce within subgroups of the entire population, isolated to some extent from one another, but with the possibility of mating occurring across the boundaries of the subgroups. A population distributed amongst a number of semi-isolated breeding groups is known as polytypic. Humans, for example, are polytypic in that they consist of groups of the species isolated from one another geographically, culturally and economically. Although breeding may occur between individuals from different subgroups of the species, it is much more likely that individuals from within the same group will reproduce together.

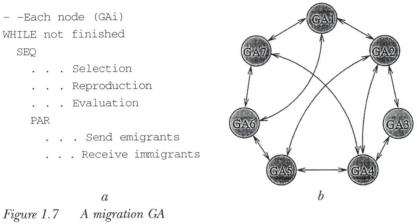

```
- -Each node (GAi)
WHILE not finished
  SEQ
     . . . Selection
     . . . Reproduction
     . . . Evaluation
  PAR
     . . . Send emigrants
     . . . Receive immigrants
```

a   *b*

Figure 1.7   *A migration GA*

       a pseudocode outline
       b a possible implementation

The migration or island model of the GA introduces geographic population distribution by dividing a large population into many smaller semi-isolated subpopulations or demes. Each subpopulation is a separate breeding unit using local selection and reproduction rules to locally evolve the species. From time to time, migration of individuals occurs between subpopulations such that individuals from one population are introduced into another subpopulation. The pattern of migration limits how much genetic diversity can occur in the global population. This pattern of migration is determined by the number of individuals migrated, the interval between migration and the migration paths between subpopulations. This movement of individuals between demes is often termed the stepping stone model.

The traditional sequential GA can readily be extended to encompass the migration model. A pseudocode outline of the modified algorithm for the migration GA is shown in Figure 1.7(a). The population is divided into a number of subpopulations each of which is evolved by an independent GA. Additional routines are included to exchange individuals between subpopulations according to the communications topology employed and global termination criteria introduced. Figure 1.7(b) shows a possible implementation of the migration GA and some of the migration paths between the population islands.

Grosso [40] first introduced a geographically isolated population structure in 1985 using an island model of the GA with five independent subpopulations. From his study, he found that semi-isolated populations improved the performance of the GA in terms of the quality of solution and the number of function evaluations required. He asserted that limited migration of individuals between subpopulations was more effective than either complete sub-population interdependence or independence.

Tanse [41] also reported on a migration GA in 1987. He studied the migration model and compared its performance with a partitioned GA, i.e. a GA whose population is divided into subpopulations which evolve entirely independently with no migration between subpopu-lations. The GA was implemented on a hypercube machine which employed custom VAX-like CPUs. This implementation used two new parameters to specify the migration interval, at which generation migration should take place, and the migration rate, the number of individuals transferred between subpopulations. In early experiments [41], Tanse selected individuals for migration probabilistically from the subset of the subpopulation whose fitness was at least equal to the average fitness of the subpopulation. Likewise, individuals were selected for replacement by immigrants probabilistically from the subset of individuals whose fitness was no greater than the average for that subpopulation. Later [42], it appears that a new strategy was adopted. At a migration generation each node produced more offspring than the current subpopulation size. The migrants were then uniformly selected from the offspring and removed from the subpopulation, thus maintaining the correct subpopulation size. The receiving subpopulation uniformly replaced individuals with immigrants. The philosophy behind this approach is that the most fit individuals are more likely to reproduce and are therefore most likely to migrate. The actual migration took place bidirectionally along one dimension of the hypercube, selected on the basis of the generation number. Thus, the neighbour selected to receive individuals from a node will also send its migrants to that node.

The results presented by Tanse showed a near-linear speed up when compared against a sequential GA with a population size equal to the sum of the individual subpopulations. Comparing the migration GA with the partitioned GA, the migration GA consistently found superior individuals and had a higher average fitness over the entire population. However, due to the limited number of test functions, no conclusion can be drawn about the general effect of the migration rate and interval. The effect of the mutation and crossover operators used with the migration GA was also investigated. The results indicate that it is feasible to use different crossover and mutation rates on different nodes, allowing the balance between exploration and exploitation to be varied locally, but with migration ensuring that good individuals should survive in at least some subpopulations.

Similar results to Tanse are reported by Starkweather *et al.* [43] and Cohoon *et al.* [44]. Starkweather's GA has a number of notable differences from the implementations described so far. Rather than using a generational GA in which most or all of the population is replaced at each generation, this parallel GA was based on Whitley's GENITOR program [35] which uses one-at-a-time reproduction replacing a single individual at each reproduction step. The migration GA was applied to a wide range of problems including neural network optimisation, a mapping problem and a 105-city travelling salesman problem. In all test cases, the migration GA produced better results than a comparable sequential one. When the migration GA was implemented on a sequential machine it was found that it would find better solutions and execute faster than a standard GA with the same population size on a number of test cases. In addition, the use of an adaptive mutation rate, initially high and reducing with generations, was found to improve the convergence characteristics of the GA.

In 1991, Mühlenbein *et al.* [45] described a real-valued parallel GA for use as a black-box solver in high-dimensional optimisation problems. A conventional single-point crossover operator was employed that operated directly on the ANSI-IEEE floating-point representation of the decision variables. The mutation operator worked only on the fractional part of a variable's representation; thus mutation is exponentially biased towards producing a new individual in the region of the original rather than one a larger numerical distance away. In experiments, the parallel GA was able to find global solutions to problems of up to a dimension of 400.

The distributed breeder GA of Mühlenbein and Schlierkamp-Voosen [26], rather than modelling the natural and self-organised evolution of the earlier parallel GA described above, is based on a model of rational selection in human breeding groups. Whereas the parallel GA models natural selection, the breeder GA models artificial selection. Using influences from evolution strategies [3] and GAs, the breeder GA selects the best T % of a population, where T is a predefined parameter, and randomly mates them until sufficient offspring are produced. As well as ensuring that no individuals mate with themselves, the fittest individual also survives into the following generation. This selection and reproduction process is known as truncation selection as only a subset of each generation is used as potential parents.

The breeder GA operates on populations of real-valued individuals and has new genetic operators designed specifically for this representation, such as intermediate and line recombination, which have been described earlier. The parallel GA uses a local hill-climbing algorithm on certain individuals to improve a current local estimate. In the breeder GA, the mutation operator was found to be almost as effective as local hill climbing but was much less complex and computationally demanding to implement. In all cases, the breeder GA was found to be more effective than the earlier parallel GA and managed to solve numerical optimisation problems of dimension 1000.

Clearly, the migration model of the GA is well suited to parallel implementation on MIMD machines. Given the range of possible population topologies and migration paths between them, efficient communications networks should be possible on most parallel architectures from small multiprocessor platforms to clusters of networked workstations. The semi-isolation of subpopulations and limited communication between them also encourage a high degree of fault tolerance. In a well designed migration GA, in the event of the loss of individual subpopulations or communications paths between them, the GA can still perform useful computation.

The migration GA has generally been reported as a more efficient search and optimisation method than conventional sequential GAs. From the preceding text, it should be clear that this is the effect of local selection and migration rather than parallel implementation. However, the migration GA is slightly more complex to use as further parameters are introduced to control migration between subpopulations.

1.4.3 Diffusion GAs

An alternative model of a distributed population structure is provided by the diffusion GA. Whereas migration introduces discontinuities into the population structure with barriers between the borders of the islands containing the subpopulations, diffusion treats the population as a single continuous structure. Each individual is assigned a geographic location on the population surface and is allowed to breed with individuals contained in a small local neighbourhood. This neighbourhood is usually chosen from immediately adjacent individuals on the population surface and is motivated by the practical communication restrictions of parallel computers. The diffusion GA is also known as the neighbourhood, cellular or fine grained GA.

Figure 1.8 shows a pseudocode outline of the diffusion GA. Consider the population distribution shown in Figure 1.9(a) where each individual, $I_{j,k}$, is assigned a separate node on a toroidal-mesh parallel processing network. The Figure shows that there are no specific islands in the population structure, rather a contiguous geographic distribution of individuals; however, there is potential for a similar effect. Given that mating is restricted to adjacent processors, then individuals on distant processors may take as many generations to meet and mate as individuals in different subpopulations in the island model. Wright [46] refers to this form of isolation within a species as isolation by distance. From Figure 1.8, each individual is first initialised, either randomly or using a heuristic, and its performance evaluated. Each node then sends its individual to its neighbours and receives individuals from those neighbours. For example, in Figure 1.9(a), individuals $I_{3,1}$ sends a copy of itself to $I_{2,1}$, $I_{3,2}$, $I_{3,5}$ and $I_{4,1}$ and receives copies of the individuals on those nodes. The purpose of this communication is to provide a pool of potential mates from the incoming individuals. Thus, selection of a mate for individual $I_{3,1}$ is made on the basis of a neighbourhood fitness over the individuals $I_{2,1}$, $I_{3,2}$, $I_{3,5}$ and $I_{4,1}$. Reproduction involves the usual crossover and mutation operators and is used to produce a single individual to replace the parent residing on the node. However, rules may be applied to retain the original parent if neither of the offspring is sufficiently fit to replace it.

At initialisation, the distribution of genetic material over the population surface is random, assuming that the population has not been seeded heuristically. After a few generations, local clusters of individuals with similar genetic material and fitness may appear in the population giving rise to virtual islands. This phenomenon is shown in Figure 1.9(b) where the shading is used to represent individuals with similar genetic material. The drift in the population caused by local selection tends to reduce the number of clusters whilst increasing their size over generations as the most fit individuals diffuse over the population.

```
— Each node (Ii,j)
WHILE not finished
SEQ
... Evaluate
PAR
            ... Send self to neighbours
            ... Receive neighbours
    ... Select mate
    ... Reproduce
```

Figure 1.8   Pseudocode outline of a diffusion GA

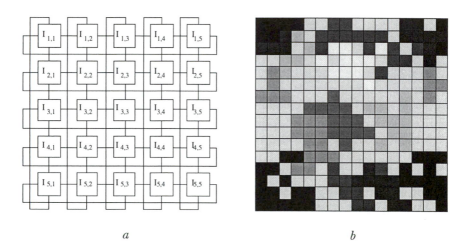

a   b

Figure 1.9   A diffusion GA
a neighbourhood communications
b virtual islands

The first attempt at a massively-parallel fine-grained GA known to the author was by Robertson [47] in 1987. Robertson used a SIMD connection machine and assigned one individual per processor. However, global selection and recombination was performed on the host machine and the individual processors were only used for function evaluation implementing a massive processor farm (Section 1.4.1). Even considering the large communications overhead of this scheme, the objective function evaluation was significantly large for huge speed up to be reported. By 1989 a more subtle and appropriate scheme for the connection machine was presented by Manderick and Spiessens [48] and later implemented on an AMT DAP [49].

Individuals were again placed on separate processors in a planar grid, but a local selection strategy based on a neighbourhood fitness distribution was used. This first diffusion algorithm was not only motivated by the desire to use the connection machine's architecture more effectively, but also to align the GA more closely with natural biological evolution. Spiessens and Manderick argue that in nature there is no global selection or fitness distribution. Rather, natural selection is a local phenomenon where individuals find a mate in their local environment. Their implementation is similar to that described here with the exception that one parent is chosen from the local neighbourhood probabilistically on the basis of the neighbourhood fitness function and recombined with a randomly selected mate within the same locality. This implementation was tested on the De Jong test-bed functions and compared with a conventional GA. The results indicate that the lower selective pressure, due to the local selection mechanism, encourages greater exploration of the search space and helps inhibit the early domination of the population by good individuals. The results also show that the parallel GA is more effective when the objective function is multimodal.

Davidor [50] in his version of the diffusion GA, called ECO GA, used a 2D grid with wraparound to produce a population surface in which each individual had eight neighbours. He used a one-at-a-time reproduction strategy and allowed the offspring produced by a particular neighbourhood to replace probabilistically an individual in the vicinity of its parents. Davidor also described the phenomena of niche and speciation where the virtual islands on the population surface represent near local optima.

An interesting variation on local selection is given by Collins and Jefferson [51]. Instead of selecting a mate from within a small local neighbourhood, an individual takes a random walk and selects a mate from individuals encountered on the way. This was found to be highly efficient in the graph partitioning problem and demonstrated a capability of finding multiple optima in a single population. In addition, four metrics were used to measure the differences in evolutionary dynamics between polytypic and panmictic populations. They were the diversity of alleles and genotypes, an inbreeding coefficient measuring the similarity between parents and speed and robustness.

More recently a number of researchers have focused on the nature of the population structure and its effect on the diffusion and convergence characteristics of the GA. For example, Baluja [52] reports on a comparative study of neighbourhood topologies. Three topologies are considered: a linear neighbourhood, a two-dimensional toroidal array and a linear neighbourhood with a rightward discontinuity. When tested on a wide range of test-bed problems the toroidal array neighbourhood consistently outperformed the other population structures. However, for some problems the linear neighbourhoods were found to produce the best convergence pattern. Baluja argues that these results are due to a combination of the effect of genetic mobility and total population size. In particular, the parameters of the different implementations, such as crossover and mutation rates, were held constant and not tuned to a particular neighbourhood structure. However, in an earlier study, Georges-Schleuter [53] observed that as the borders in one-dimensional population structures are smaller than those in two dimensions, local niches once established tend to survive for longer periods.

The diffusion model provides a finer grain of parallelism than that of the worker/farmer and island models. It is suitable for implementation on a wide range of parallel architectures from single-bit digital array processors (DAPs) and massively parallel SIMD machines like the connection machine, to MIMD computers, such as transputer networks. There are even reports in the literature of fine-grained GAs being implemented on clusters of networked workstations [54].

The basic operator to support the diffusion model is that of a local neighbourhood selection mechanism. From the examples of diffusion models presented in this Section it is clear that this local selection results in performance superior to that of global and migration GAs with comparable population sizes. Good solutions are found faster, requiring fewer function evaluations, and different solution niches may be established in the same evolutionary cycle. Furthermore, diffusion appears to implement a more robust search in the presence of deceptive or GA-hard objective functions.

1.5 GAs for engineering systems

As the GA does not require derivative information or a formal initial estimate of the solution region and because of the stochastic nature of the search mechanism, GAs are capable of searching the entire solution space with more likelihood of finding the global optimum than conventional optimisation methods. Indeed, conventional methods usually require the objective function to be well behaved, whereas the generational nature of GAs can tolerate noisy and discontinuous function evaluations. For these and other reasons, the engineering community has been quick to see the potential of GAs.

A number of considerations commonly arising in control engineering problems, and the way in which these are treated through the application of GAs are discussed below.

Representation: Continuous decision variables may be handled either directly through real-valued representations and the appropriate genetic operators or by using binary representation schemes and standard genetic operators. In the case of binary representations, real values can be approximated to the necessary degree with a fixed-point binary scheme. In most engineering problems, however, it is the relative precision of the parameters that is significant rather than absolute precision. In such cases, the logarithm of the parameter may be encoded reducing the number of bits and hence memory usage. Alternatively, a direct floating-point representation may be used. Discrete decision variables can be encoded using either binary or *n*-ary representation. When functions can be expected to be locally monotonic with respect to such variables, the use of Gray coding is known to better exploit that monotonicity. This consideration also applies to binary representations of continuous decision variables. In cases where a mixture of discrete and continuous decision variables is to be used, it is feasible to use a mixed representation provided that care is taken to ensure that the genetic operators used function correctly over the set of encodings chosen. However, encoding all of the parameters using a single binary representation can simplify the operation of the EA.

Scale: The concept of fitness is central to all EA approaches. Given that many optimisation problems are characterised by a real-valued objective function, these values must be converted into a non-negative fitness value if they are to be handled correctly by the EA. Early work on GAs concentrated on the use of offsetting objective function values so that selection could be based directly on an individual's performances within a population [6]. The use of scaling retains an individual's relative performance and also attempts to bias the selective pressure towards better individuals although still allowing relatively unfit individuals the potential to reproduce. Alternatively, by discarding the relative differences between individuals' raw performances and only considering them on their rank in a population, a constant selective pressure may be applied throughout the evolutionary process. Offsetting and scaling can result in more and more individuals receiving fitnesses with relatively small differences as the population converges. The rank-based methods maintain a constant selective pressure towards good individuals throughout the convergence process and are claimed to bring a number of other advantages [35]. Additionally, rank-based schemes also offer a convenient mechanism for considering multiple objectives simultaneously and this is discussed in Chapter 3.

Constraints: Most engineering problems are subject to constraints. For example, actuators have finite limits on the loads which can move and positions they can reach and control loops are usually required to be stable. EAs can handle constraints in a number of ways. The most efficient and direct method is to embed these constraints in the coding of the individuals. Where this is not possible, penalty functions may be used to ensure that invalid individuals have fitnesses which reflect that they are low performers. However, appropriate penalty functions are not always easy to design for a given problem and may affect the efficiency of the search [55]. An alternative approach is to consider constraints as design objectives and recast the problem as a multiobjective one. Again, this is discussed in Chapter 3.

Adaptation: The vast majority of applications of EAs have concentrated on their use as a function optimiser. However, EAs have been shown to be well suited to tracking time-varying systems [56], i.e. ones in which the optimum fitness or fitness criterion changes over time. Such changes typically occur as a result of a change in the external environment, e.g. a change in operating conditions, or due to system changes, e.g. wear of mechanical components. The EA has the advantage over many conventional methods of being able to respond to such changes by exploiting the diversity of the individuals in the current population. If there is insufficient diversity in the population, then new material can be readily introduced by replacing some individuals with randomly initialised individuals.

Software: Although there exist many good public-domain genetic algorithm packages, such as GENESYS [57] and GENITOR [35], none of these provides an environment which is immediately compatible with existing tools in the control domain. The MATLAB Genetic Algorithm Toolbox [58] aims to make GAs accessible to the control engineer within the framework of an existing CACSD package. This allows the retention of existing modelling and simulation tools for building objective functions and allows the user to make direct comparisons between genetic methods and traditional procedures. By building the EAs on a standard computer aided control system design

(CACSD) package, EAs can be made available to control engineers as a powerful tool to complement those already in use. The NeuralWorks Professional II/Plus neural network software from NeuralWare Inc. now comes with a genetic reinforcement learning system that augments the standard training procedures using an EA to avoid getting stuck in local optima and it can be expected that many CACSD and CAE packages will have EA tools available in the near future.

1.6 Example application: gas turbine engine control

From the preceding Sections, it is clear that the GA is substantially different from conventional enumerative and calculus-based search and optimisation techniques. In this Section, an example is presented which demonstrates how the GA may be used to address a problem that is not amenable to efficient solution via these conventional methods. The problem is to find a set of control loops and associated controller parameters for an aircraft gas turbine engine control system to meet a number of conflicting design criteria [59].

The mechanical layout of a typical twin spool gas turbine engine is shown in Figure 1.10. Each spool comprises a number of compressor and turbine stages and is aerothermodynamically coupled to the other. Air is drawn into the fan (or LP compressor) through the inlet guide vanes, which are used to match the airflow to the fan characteristics, and compressed. The air is then further compressed by the HP compressor before being mixed with fuel and combusted and then expelled through the HP and LP turbines. A portion of the air from the fan exit may bypass the HP compressor and turbines and be mixed with the combusted air/fuel mixture before being ejected through the jet pipe and nozzle to produce thrust.

The characteristics of operation of a fixed cycle gas turbine engine, such as specific thrust and specific fuel consumption, are fundamental to the engine design. The design thus becomes a compromise between meeting the conflicting requirements for performance at different points in the flight envelope and the achievement of low life-cycle costs, while maintaining structural integrity. However, variable geometry components, such as the inlet guide vanes (IGV) and nozzle area (NOZZ), may be used to optimise the engine cycle over a range of flight conditions with regard to thrust, specific fuel consumption and engine life, assisting in the reduction of life-cycle costs [60].

Dry engine control of a conventional gas turbine engine is normally based on a single closed-loop control of fuel flow (WFE) for thrust rating, engine idle and maximum limitary and acceleration control. The closed-loop concept provides accuracy and repeatability of control of defined engine parameters under all operating conditions, and compensates automatically for the effects of engine and fuel system ageing.

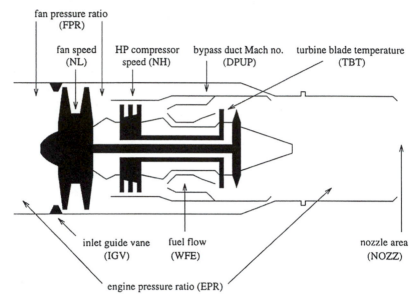

Figure 1.10   A gas turbine engine

It is usual for any variable geometry in these engines to be positioned according to commands scheduled against appropriate engine and/or aircraft parameters. These schedules often need to be complex functions of several parameters, and adjustments may be required frequently to achieve the desired performance. Clearly, success of this open-loop mode of control is reliant on the positional accuracy achievable as there is no self trimming to account for ageing as in closed-loop modes. This results in penalties of reduced engine life and higher maintenance costs.

1.6.1 Problem specification

The object of the design problem is to select a set of sensors and design a suitable controller for a manoeuvre about a particular operating point while meeting a set of strict performance criteria. Figure 1.11 shows the basic simulation model used for this example. A linearised model of the Rolls-Royce Spey engine, with inputs for fuel flow, exhaust nozzle area and HP inlet guide vane angle, is used to simulate the dynamic behaviour of the engine. Sensors provided from outputs of the Spey engine model are high and low pressure spool speed (NH and NL), engine and fan pressure ratios (EPR and FPR) and bypass duct Mach number (DPUP). These sensed variables can be used to provide closed-loop control of WFE and NOZZ. Three inputs are provided; input one is the demand reference signal and is translated by the power lever angle (PLA) resolver to provide the reference signal for the fuel flow control loop, and inputs two and three determine the measured parameters used to provide closed loop control. In this example, the nozzle area demand signal is derived from the fan working line, and positioning of the HP IGVs is directly scheduled against the HP spool speed. The possible control loops are thus:

WFE	NL
	NH
	EPR
NOZZ	open-loop schedule
	FPR
	DPUP

giving nine possible controller configurations.

   For simplicity, each loop is to be controlled by a PI controller and a single 50% thrust-rating operating point considered at sea level static conditions. The system is required to meet the following design objectives:

1   70% rise-time ≤ 1.0 s
2   10% settling time ≤ 1.4 s
3   XGN ≥ 40 KN
4   TBT ≤ 1540 K
5   LPSM ≥ 10 %
6   WFE sensitivity ≤ 2 %
7   $\gamma \leq 1$

where objectives 1 and 2 are in response to a change in thrust demand of 33.33% to 66.66%, XGN is the engine gross thrust, TBT the maximum turbine blade temperature, LPSM the fan surge margin and the WFE sensitivity is a result of a one per cent error in the sensed control parameter. Additionally, the system should be closed-loop stable (objective 7).

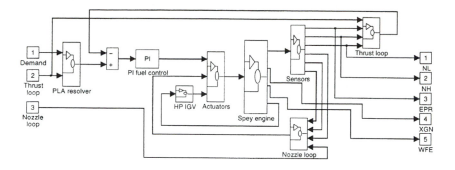

Figure 1.11   Simulation model

1.6.2 EA implementation

The basic simulation model was developed in the MATLAB/SIMULINK CACSD package and the associated performance measurements determined by simulation. Additionally, further models were constructed for the sensitivity and stability objectives. Actuators were modelled as first order systems with appropriate time constants and sensor parameters derived from the linear engine model outputs. To include realistic acceleration protection, input demands were rate limited.

A structured chromosome representation [56] was employed to allow the controller parameters for all possible control loops to reside in all individuals, Figure 1.12. Here, high-level genes, labelled WFE and NOZZ, are used to determine which control loops, and hence sensors, of Figure 1.11 are used. Associated with each control loop are the parameters for the corresponding PI controller, $\{Pi_j, Ii_j\}$. Note that as the NOZZ loop may be open-loop scheduled, there are no $P2_0$ and $I2_0$ parameters. In this manner, the chromosome may contain a number of good representations simultaneously, although only the set defined by the high-level genes will be active.

Multiobjective ranking, fitness sharing and mating restrictions are used with standard GA routines to implement the multiobjective genetic algorithm (MOGA) [61]. Multiobjective ranking is based upon the dominance of an individual and how many individuals outperform it in objective space, combined with goal and priority information. In this example, the goals were set to the values given in Section 1.6.1 and all objectives were assigned the same priority. In cases where objectives are assigned different priorities, higher priority objectives are optimised in a Pareto fashion until their goals are met at which point the remaining objectives are optimised. Fitness sharing, implemented in the objective domain favours sparsely populated regions of the trade-off surface and may be combined with mating restrictions to reduce the production of low performance individuals by encouraging the mating of individuals similar to one another. A full discussion of MOGAs may be found in Chapter 3.

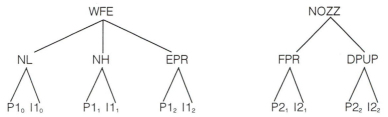

Figure 1.12   Structured chromosome representation

In the example presented here, a binary MOGA with a population of 70 individuals was employed. The integer variables for WFE and NOZZ loop selection were encoded with eight bits and each controller parameter with 16 bits. Finally, lists of nondominated solutions for each controller configuration were maintained throughout the execution of the MOGA.

1.6.3 Results

Figure 1.13 illustrates a typical trade-off graph for Spey engine controller designs, each line representing a non-dominated solution found by the MOGA. The x axis shows the design objectives, the y axis the performance of controllers in each objective domain and the crossmarks in the Figure show the design goals.

In Figure 1.13, only the preferred individuals, those that satisfy the design goals, are shown. When no individuals satisfy all the design goals, the nondominated or Pareto optimal solutions are displayed. Trade-offs between adjacent objectives result in the crossing of the lines between them whereas concurrent lines indicate that the objectives do not compete with one another. For example, the power rating and TBT (objectives 3 and 4) appear to compete quite heavily although the rise time and settling time (objectives 1 and 2) do not exhibit the same level of competition.

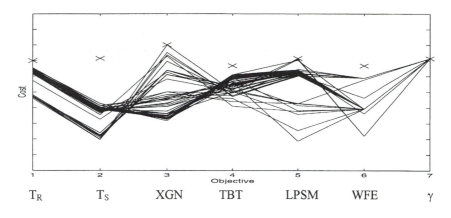

Figure 1.13   Sample trade-off graph

The information contained in such trade-off graphs can be applied in a number of ways. For example, the transient response and thrust margins can be used to help establish derates from the nominal engine model and set suitable performance commitments when nonideal components and control are employed in production. The LPSM may be seen as a measure of the stability of the controlled engine and used to assess the adequacy of the control in conjunction with the stability objective. TBT and sensitivity may be used with nominal engine data to establish mechanical design criteria and fed back into the engine design process. For example if TBT needed to be increased then the turbine design may require further refinement.

Having satisfied the design goals or established which are unattainable, the control engineer is now free to assess the relative merits of each controller design. Figure 1.14 shows trade-off graphs for the nondominated solutions found for each controller type grouped for the main fuel control loop. This data may be used to examine control, performance and other aspects of the system design.

From the trade-offs shown in Figure 1.14, it can be seen that NL control provides the best step response characteristics but suffers from a relatively high degree of sensitivity to sensor error. The sensitivity margin may be improved, for example, by selecting a better type of sensor for NL. The lowest power controllers were also found indicating that although a good thrust mapping to PLA can be achieved, there is less scope for compensation on a nonideal engine.

On the other hand, the NH controllers minimise the TBT at the expense of a slower response to changes in demand. This may allow the use of cheaper turbine materials, for example, if the other

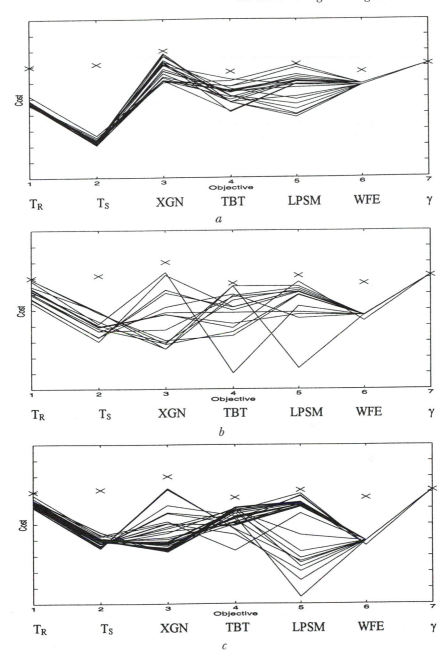

Figure 1.14   Trade-offs for WFE control loops
   a all NL controllers
   b all NH controllers
   c all EPR controllers

performance criteria are satisfactory or indicate a longer engine life expectancy and a reduced cost of ownership. Sensitivity to sensor error is better than with NL control and higher thrust ratings may also be achieved.

EPR control results in the least sensitivity to sensor error and allows a larger LPSM to be maintained. Step response times are generally the slowest although better TBT control could be achieved over NL control. Improvements in the step response could be achieved, but at the expense of reduced LPSM and increased TBT.

1.6.4 Discussion

This control system design example has illustrated an application for which the use of GAs offers a number of significant advantages over other search and optimisation methods. The problem considered contained a mixture of discrete and continuous decision variables (i.e. the control loops and controller parameters) and the GA was able to deal with them in a simple, straightforward manner. Selecting the controller gains using, say, gradient-based optimisation would require the formulation of nine different optimisation problems. If more potential closed-loop controls had been considered, or if the control structure was allowed to vary, then it is easy to imagine that even relatively trivial problems would be beyond the coverage of most other techniques.

Operating on a population of solution estimates, the GA is often able to identify a number of potential solutions to a given problem. This is particularly important for the example presented here where the problem considered was multiobjective. The use of MOGA extensions allows us to build a picture of the trade-offs and conflicts which exist between the different design objectives. Thus, we can select a final solution with greater confidence that it will satisfy all of our design criteria and is the most suitable for a particular application.

1.7 Concluding remarks

This Chapter has presented a broad survey of the current trends and techniques used in GAs. Many variations from the original GA have been discussed, such as representation and reproduction strategies, and a broad overview of parallel implementations has been given. Clearly, from the material presented in this text, it can be seen that the GA is a powerful and versatile search and optimisation method applicable to a broad range of activities.

The remainder of this book is dedicated to the practical application of GAs in engineering systems. It is hoped that this volume will convey to the reader a feeling for the wide variety of engineering applications of this versatile technique and encourage individuals to explore the potential of genetic algorithms in their own field.

1.8 References

1 Spears, W. M., De Jong, K. A., Bäck, T. , Fogel, D. B. and de Garis, H.: 'An overview of evolutionary computation'. Machine learning. ECML-93 European conference on *Machine learning*, lecture notes in artificial intelligence, No. 667, pp. 442–459, 1993

2 Fogel, L. J., Owens, A. J. and Walsh, M. J.: *Artificial intelligence through simulated evolution* (Wiley Publishing, New York, 1966)

3 Rechenberg, L.: *Evolutionsstrategie: optimierung technischer systeme nach prinzipien der biologischen evolution* (Frommann-Holzboog, Stuttgart, 1973)

4 Koza, J. R.: *Genetic programming: On the programming of computers by means of natural selection* (MIT Press, Cambridge, Massachusetts, 1992)

5 Holland, J.: *Adaptation in natural and artificial systems* (University of Michigan Press, 1975)

6 Goldberg, D. E.: *Genetic algorithms in search, optimisation and machine learning* (Addison Wesley Publishing Company, January 1989)

7 Karr, C. L.: Design of an Adaptive Fuzzy Logic Controller Using a Genetic Algorithm. Proc. 4th int. conf. on *Genetic algorithms*, pp. 450–457, 1991

8 Holstien, R. B.: *Artificial genetic adaptation in computer control systems*. PhD thesis, Department of Computer and Communication Sciences, University of Michigan, Ann Arbor, 1971

9 Caruana, R. A., and Schaffer, J. D.: Representation and Hidden Bias: Gray vs. Binary Coding. Proc. 6th int. conf. *Machine learning*, pp.153–161, 1988

10 Schmitendorgf, W. E., Shaw, O., Benson R. and Forrest, S.: 'Using genetic algorithms for controller design: simultaneous stabilisation and eigenvalue placement in a region'. Technical report no. CS92-9, Dept. Computer Science, College of Engineering, University of New Mexico, 1992

11 Bramlette, M. F.: 'Initialization, mutation and selection methods in genetic algorithms for function optimisation'. Proc. 4th int. conf. on *Genetic algorithms*, pp. 100–107, 1991

12 Lucasius, C. B., and Kateman, G.: 'Towards solving subset selection

problems with the aid of the genetic algorithm', in *Parallel problem solving from nature* 2, Männer, R.   and Manderick, B. (eds.), Amsterdam: North-Holland, 1992), pp. 239–247

13 Wright, A. H.: 'Genetic algorithms for real parameter optimisation', in *Foundations of genetic algorithms*, Rawlins, J. E. (ed.) (Morgan Kaufmann, 1991), pp. 205–218

14 Michalewicz, Z.: *Genetic algorithms + data structures = evolution programs* (Springer Verlag, 1992)

15 Back, T., Hoffmeister, F., and Schwefel, H.-P.: 'A survey of evolution strategies'. Proc. 4th int. conf. on *Genetic algorithms*, pp. 2–10, 1991

16 Grefenstette, J. J.: 'Incorporating problem specific knowledge into genetic algorithms', in *Genetic algorithms and simulated annealing*, Davis, L. (ed.) (Morgan Kaufmann, 1987)

17 Whitley, D., Mathias, K. and Fitzhorn, P.: 'Delta coding: an iterative search strategy for genetic algorithms'. Proc. 4th int. conf. on *Genetic algorithms*, pp. 77–84, 1991

18 De Jong, K. A.: *Analysis of the behaviour of a class of genetic adaptive systems.* PhD thesis, Dept. of Computer and Communication Sciences, University of Michigan, Ann Arbor, 1975

19 Baker, J. E.: 'Adaptive selection methods for genetic algorithms'. Proc. 1st int. conf. on *Genetic algorithms*, pp. 101–111, 1985

20 Baker, J. E.: 'Reducing bias and inefficiency in the selection algorithm'. Proc. 2nd int. conf. on *Genetic algorithms*, pp. 14–21, 1987

21 Booker, L.: 'Improving search in genetic algorithms', in *Genetic algorithms and simulated annealing*, Davis, L. (ed.) (Morgan Kaufmann Publishers, 1987), pp. 61–73.

22 Spears, W. M., and De Jong, K. A.: 'An analysis of multi-point crossover', in *Foundations of genetic algorithms*, Rawlins, J. E. (ed.) (Morgan Kaufmann, 1991) pp. 301–315

23 Syswerda, G.: 'Uniform crossover in genetic algorithms'. Proc. 3rd int. conf. on *Genetic algorithms*, pp. 2–9, 1989

24 Spears, W. M., and De Jong, K. A.: 'On the virtues of parameterised uniform crossover'. Proc. 4th int. conf. on *Genetic algorithms*, pp.230–236, 1991

25 Caruana, R. A., Eshelman, L. A., and Schaffer, J. D.: 'Representation and hidden bias II: eliminating defining length bias in genetic search via shuffle crossover', in *Eleventh international joint conference on artificial intelligence*, Sridharan, N. S. (ed.) (Morgan Kaufmann, 1989), vol. 1, pp. 750–755

26 Muhlenbein, H., and Schlierkamp-Voosen, D.: 'Predictive Models for the Breeder Genetic Algorithm', *Evolutionary Computation*, **1**, (1), pp. 25–49, 1993

27 Furuya, H., and Haftka, R. T.: 'Genetic algorithms for placing actuators on space structures'. Proc. 5th int. conf. on *Genetic algorithms*, pp. 536–542, 1993

28 Falkener, E.: 'A new representation and operators for genetic algorithms applied to grouping problems', *Evolutionary Computation*, **2**, (2), pp. 123–144, 1994

29 Baxter, M. J., Tokhi M. O. and Fleming, P. J.: 'An investigation of the heterogeneous mapping problem using genetic algorithms'. Proc. UKACC *Control '96*, **1**, pp. 448-453, 1996

30 Janikow, C. Z., and Michalewicz, Z.: 'An experimental comparison of binary and floating point representations in genetic algorithms', Proc. 4th int. conf. on *Genetic algorithms*, pp. 31–36, 1991

31 Tate, D. M., and Smith, A. E.: 'Expected allele convergence and the role of mutation in genetic algorithms', Proc 5th int. conf. on *Genetic algorithms*, pp.31–37, 1993

32 Davis, L.: 'Adapting operator probabilities in genetic algorithms'. Proc. 3rd int. conf. on *Genetic algorithms*, pp. 61–69, 1989

33 Fogarty, T. C.: 'Varying the probability of mutation in the genetic algorithm'. Proc. 3rd int. conf. on *Genetic algorithms*, pp. 104-109, 1989

34 De Jong, K. A., and Sarma, J.: 'Generation gaps revisited', in *Foundations of genetic algorithms* 2, Whitley, L. D. (ed.), (Morgan Kaufmann, 1993)

35 Whitley, D.: 'The GENITOR algorithm and selection pressure: why rank-based allocations of reproductive trials is best'. Proc. 3rd int. conf. on *Genetic algorithms*, pp. 116–121, 1989

36 Huang, R., and Fogarty, T. C.: 'Adaptive classification and control-rule optimisation via a learning algorithm for controlling a dynamic system', Proc. 30th conf. *Decision and control*, Brighton, England, pp. 867–868, 1991

37 Fogarty, T. C.: 'An incremental genetic algorithm for real-time learning'. Proc. 6th int. workshop on *Machine learning*, pp. 416–419, 1989

38 Schwefel, H. -P.: *Numerical optimisation of computer models*, (John Wiley and Sons, New York, 1981)

39 Dodd, N., Macfarlane, D., and Marland, C.: 'Optimisation of artificial neural network structure using genetic techniques implemented on multiple transputers', in *Transputing '91*, Welch, P., Stiles, D., Kunii, T. L. and Bakkers, A. (eds.) (IOS Press, 1991), vol. 2, pp. 687–700

40 Grosso, P. B.: *Computer simulation of genetic adaptation: parallel subcomponent interaction in a multilocus model.* PhD thesis, University of Michigan, 1985

41 Tanse, R., 'Parallel genetic algorithm for a hypercube'. Proc. 2nd int. conf. on *Genetic algorithms*, pp. 177–183, 1987

42 Tanse, R.: 'Distributed genetic algorithms'. Proc. 3rd int. conf. on *Genetic algorithms*, pp.434–439, 1989

43 Starkweather, T., Whitley, D. and Mathias, K., 'Optimisation using distributed genetic algorithms", in *Proc. Parallel Problem Solving From Nature 1, Lecture Notes in Computer Science No. 496* (Springer-Verlag, 1990) pp. 176–185

44 Cohoon, J. P., Martin, W. N. and Richards, D. S.: 'A multi-population genetic algorithm for solving the K-partition problem on hypercubes'. Proc. 4th int. conf. on *Genetic algorithms*, pp. 244-248, 1991

45 Mühlenbein, H., Schomisch M. and Born, J. 'The parallel genetic algorithm as a function optimizer', *Parallel Comput.* (17), pp. 619–632, 1991

46 Wright, S.: *Evolution and the genetics of populations* (University of Chicago Press, 1969), vol. 2

47 Robertson, G.: 'Parallel implementation of genetic algorithms in a classifier system', in *Genetic algorithms and simulated annealing*, Davis, L. (ed.) (Pitman, London, 1987), pp. 129–140

48 Manderick, B., and Spiessens, P. 'Fine-grained parallel genetic algorithms'. Proc. 3rd int. conf. on *Genetic algorithms*, pp. 428–433, 1989

49 Spiessens P., and Manderick, B.: 'A massively parallel genetic algorithm: implementation and first analysis'. Proc. 4th int. conf. on *Genetic algorithms*, pp. 279–286, 1991

50 Davidor, Y.: 'A naturally occurring niche and species phenomenon: the model and first results'. Proc. 4th int. conf. on *Genetic algorithms*, pp. 257–263, 1991

51 Collins, R. J., and Jefferson, D. R.: 'Selection in massively parallel genetic algorithms'. Proc. 4th int. conf. on *Genetic algorithms*, pp. 249–256, 1991

52 Baluja, S.: 'Structure and performance of fine-grain parallelism in genetic search', Proc. 5th int. conf. on *Genetic algorithms*, pp. 155–162, 1993

53 Georges-Schleuter, M.: 'Comparison of local mating strategies in massively parallel genetic algorithms', in *Parallel problem solving from nature 2*, Männer, R. and Manderick, B. (eds.), (Amsterdam: North-Holland, 1992), pp. 553–562)

54 Maruyama, T., Hirose, T. and Konagaya, A.: 'A fine-grained parallel genetic algorithm for distributed parallel systems'. Proc. 5th int. conf. on *Genetic algorithms*, pp. 184-190, 1993

55 Richardson, J. T., Palmer, M. R., Liepins, G., and Hilliard, M.: 'Some guidelines for genetic algorithms with penalty functions'. Proc. 3rd int. conf. on *Genetic algorithms,* pp. 191-197, 1989

56 Dasgupta D., and McGregor, D. R.: 'Nonstationary function optimisation using the structured genetic algorithm', in *Parallel problem solving from nature 2,* Männer, R. and Manderick, B., (eds.), (Amsterdam: North-Holland, 1992) pp. 145-154.

57 Grefenstette, J. J.: 'A user's guide to GENESIS version 5.0', *Technical Report,* Navy Centre for Applied Research in Artificial Intelligence, Washington D.C., USA, 1990

58 Chipperfield, A. J., Fleming, P. J. and Fonseca, C. M.: 'Genetic algorithm tools for control systems engineering'. Proc. 1st int. conf. on *Adaptive computing in engineering design and control,* Plymouth Engineering Design Centre, UK, 21-22 September, pp. 128-133, 1994

59 Chipperfield A. J. and Fleming, P. J.: 'Systems integration using evolutionary algorithms'. Proc. UKACC *Control '96,* **1,** pp. 705-710, Exeter, UK, 1996

60 Garwood, K. R. and Baldwin, D. R. 'The emerging requirements for dual and variable cycle engines', 10th int. symp. on Air *breathing engines,* RAE-TMP-1220, 1992

61 Fonseca, C. M., and Fleming, P. J.: 'An overview of evolutionary algorithms in multiobjective optimisation', *Evolutionary Computing,* **3,** (1), pp. 1-16, 1995

Chapter 2
Levels of evolution for control systems

J. J. Grefenstette

2.1 Introduction

Evolutionary algorithms (EAs) are general purpose search and learning methods that can be applied to a variety of problems relating to control systems. This Chapter focuses on the range of representation levels at which evolutionary algorithms can be applied to control systems, including evolving control parameters, evolving complex control structures and evolving control rules. The discussion also outlines the use of evolutionary algorithms for testing intelligent control systems. In this case, the EA is used to identify weaknesses in a control system by searching for challenging test cases.

2.1.1 Evolutionary algorithms

Evolutionary algorithms are heuristic search and learning algorithms based on principles derived from biological evolution. Although many variations exist, an outline of the general idea is shown in Figure 2.1. One key aspect of evolutionary algorithms is that, unlike most traditional search methods, EAs maintain a population of candidate solutions rather than a single candidate. The population evolves through a process of (1) selecting the more highly fit individuals for replication, (2) mutating these individuals to create new alternative candidate solutions and (3) recombining individuals with other individuals to create new combinations of features.

During the past two decades, many variations of this general approach have been studied. Particular styles of EAs are known by such names as genetic algorithms (GA), evolution strategies (ES) and evolutionary programming (EP). Although the GA community initially focused on simple string representations which support general purpose problem solving and binary recombination operators like crossover, the ES community has generally focused on engineering applications involving function optimisation, relying on sophisticated mutation operators to create variants in the population.

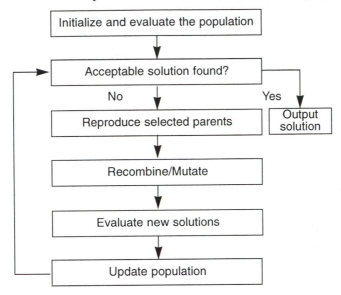

Figure 2.1 Outline of an evolutionary algorithm

Evolutionary programming focuses on the evolution of finite-state models for prediction and control.

Some specific dimensions along which EAs differ are:

- How individual candidate solutions are represented: e.g., bit strings, real-valued vectors, rules.
- How the population is updated: e.g., generational or incremental updates.
- How individuals are selected for reproduction: e.g., proportional selection, rank-based selection, tournament selection, threshold selection, (μ, λ) and $(\mu + \lambda)$ selection.
- How parent individuals are mated: e.g., n-point, uniform, arithmetic crossover.
- How individuals are mutated: e.g., bit flipping, hill climbing, knowledge-based changes.
- How other parameters of the algorithm are set: e.g., population size, operator rates.

Some of these topics are covered in Chaper 1 of this volume. For a detailed discussion of these issues in evolutionary algorithms, see [1]. We now turn to some general issues arising in EA applications in control.

2.1.2 Control system applications

Evolutionary algorithms offer a general purpose tool which can be used across the entire spectrum of control system applications, some of which are illustrated in Figure 2.2. Applications to control systems include control system parametric design [9, 3, 7, 26], system identification [17, 25], systems integration [2] and adaptive control [20]. See [4] for a recent survey of applications of evolutionary algorithms to control systems engineering.

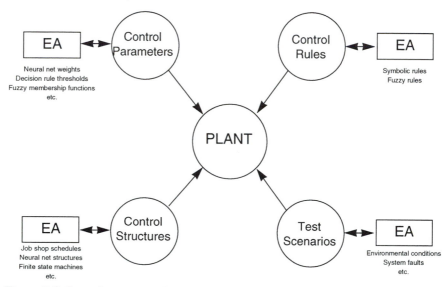

Figure 2.2 Control system applications of evolutionary algorithms

Before applying an evolutionary algorithm to any problem, two requirements must be met. First, a performance measure must be available to serve as the basis for the fitness calculation by the EA. For a control system, the performance measure might be a function of the stability of the system, the efficiency of the system, or some other measure. Although it is possible to have a fitness function based on human judgement, the efficiency of the search will be greatly enhanced if the fitness function is computed by an automated process. In some cases, developing a suitable automated fitness function for a real control system is difficult. For example, an intelligent control system for an autonomous vehicle might be required to diagnose internal system faults and respond appropriately. Developing an automated critic for such a system presents a significant challenge.

Secondly, it must be feasible to compute many evaluations of the candidate solutions. During its run, the EA will typically consider hundreds or thousands of candidate solutions. For many real control systems, e.g., power plants, experimenting with many alternative settings for control variables may be very costly, time consuming or dangerous. Therefore, evolutionary algorithms are usually applied to a system model rather than to a running control system. Ideally, the system model captures all the relevant features of the actual system that are needed to compute an accurate performance measurement. It is also desirable that the system model be much faster to evaluate than the actual system. Unfortunately, these requirements often tend to conflict in practice. Nonetheless, advances in digital simulation technology are likely to make it feasible to create increasingly more realistic models of complex control systems in the future.

2.1.3 Overview

Most applications of evolutionary algorithms deal with the optimisation of a set of parameters. However, the basic principles upon which evolutionary systems operate, namely, survival of the fittest and inheritance with variation, can be applied to a much richer set of representations. The remainder of this Chapter emphasises the different levels of representation that an EA might use to optimise a control system, including:

- evolving numeric control parameters
- evolving complex control structures
- evolving control rules.

These approaches are not mutually exclusive, since a given EA might address a combination of these levels. The intent here is to suggest to the potential user a broader range of applications than parameter optimisation alone.

After a brief consideration of the first two levels, the focus will be on the approach to evolving control rules embodied in the SAMUEL system [13]. The discussion then turns to use of evolutionary algorithms for testing intelligent control systems. In this case, the EA is used to identify weaknesses in a control system by searching for challenging test cases.

2.2 Evolutionary learning: parameters

Many control system applications of evolutionary algorithms address the problem of tuning numeric parameters [1, 2, 3, 9, 10, 15, 20, 26], as shown in Figure 2.3. The process to be controlled consists of a black box with a (perhaps large) set of control parameters. These parameters serve as control knobs that must be set in the right configuration to optimise the performance of the process. In many cases of interest, there are complex and poorly understood interactions among the control parameters, so finding an optimal or near-optimal setting presents a difficult search problem.

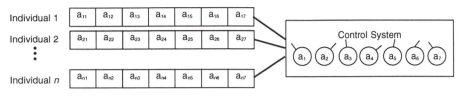

Figure 2.3   Evolving control parameters

An evolutionary algorithm can be applied to this problem in a relatively straightforward way by letting each individual specify a setting for each control parameter. Parameter settings may be represented either by bit strings or real-valued alleles. Individuals are evaluated by running the control process using the specified settings and measuring its performance. Mutation consists of altering individual parameter settings. Recombination involves combining parameters taken from different parents.

This approach to evolving control parameters can be applied to a range of rather distinct underlying control systems. In traditional three-term PID controllers, the control parameters may include various gain parameters in the control law [20]. Genetic algorithms have also been used to learn the parameters for dynamic neural networks for robotic control [28] and the control parameters for a reactive robotic navigation system [18]. The following example typifies the use of evolutionary algorithms for tuning control parameters.

In [15], the control system consists of a fuzzy logic controller (FLC). An FLC uses fuzzy rules such as:

if c_1 is LARGE and c_2 is SMALL then a_k is MEDIUM

where c_i is a condition variable and a_j is an action variable. The linguistic terms SMALL, MEDIUM and LARGE may be defined separately for each condition and action variable using fuzzy membership functions, as shown in Figure 2.4. Fuzzy membership functions are usually determined by a few parameters, e.g., the endpoints (LB and UB in Figure 2.4) of the range of condition variable values which match the linguistic variable, with the highest membership value corresponding to the centre of that range.

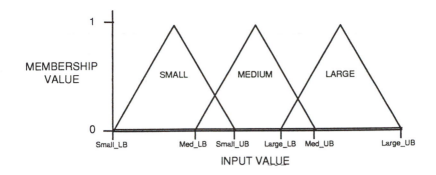

Figure 2.4   Fuzzy membership functions

In Reference 15 the fuzzy control rules in the FLC are specified in advance. A genetic algorithm is used to search for parameters of the membership functions of the linguistic variables which yield high-performance fuzzy control laws, using the representation shown in Figure 2.5.

| Individual *i:* | ••• | Small_LB | Small _UB | Med_LB | Med_UB | Large_LB | Large_UB | ••• |

Figure 2.5   Representation of fuzzy membership parameters

This volume contains several other applications of EAs to the parameter optimisation problem. These examples illustrate that evolutionary algorithms can be useful for optimising the parameters of an impressively wide class of control architectures.

2.3 Evolutionary learning: data structures

In some cases, the control process may depend on data structures that are more complex than simple lists of numeric parameters. A factory control system may require a schedule which assigns jobs to machines. A delivery control system may require a cost-efficient route to a large number of delivery sites. Evolutionary algorithms have been widely applied to such combinatorial optimisation problems in which the task is to search a space of permutations or other complex data objects, as shown in Figure 2.6. In this case, each individual in the EA's population specifies the data structure (e.g., schedule, route, etc.) which serves as input to the control system. As before, individuals are evaluated by running the control process (or a simulation) using the

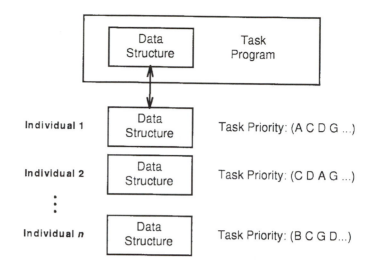

Figure 2.6   Evolving control structures

specified data structures and evaluating its performance.

An important consideration in this approach is that the genetic operators, such as mutation and recombination, need to be defined appropriately for the chosen representation. The classical genetic operators of random mutation and cut and splice crossover may not yield meaningful results on representations of permutations. For example, if the permutations:

(A B C D E F) and (A C E F B D)

undergo one-point crossover after the third position, the results are:

(A B C F B D) and (A C E D E F)

In this case, neither offspring is a legal permutation. Numerous specialised crossover operators have been developed to deal with order-based representations [8], with the goals of guaranteeing that offspring always represent legal structures, but still inherit as much information as possible from their parents. Recent examples of using genetic algorithms to evolve complex control structures include References 6 and 27.

2.4 Evolutionary learning: program level

It may be desirable to give an evolutionary algorithm even more flexibility to alter the control rules of a system. Figure 2.7 illustrates one approach using a genetic algorithm to learn control rules. Each individual in the population consists of a set of rules. Individuals are evaluated by inserting the rules into the control program (or simulation model of the control program) and measuring the performance of the resulting controller over some period of time. Using this measurement as a fitness value, the genetic algorithm selects high-performance rule sets for replication, mutation and recombination. Mutation of rules consists of altering the conditions, the actions or both. Recombination consists of exchanging rules between selected parents. Over time, clusters of rules that are associated with consistently high performance can be expected to spread throughout the population, yielding an improved set of control rules.

There are several variations on this general approach. Genetic algorithms operating on pattern-matching production systems were developed by Smith [24]. In genetic classifier systems [14], the individuals each consist of a single rule and the control program consists of the entire population of rules. In genetic programming [16], each individual consists of a tree-structured expression in a suitable language (e.g., LISP) and crossover consists of the exchange of subtrees between parent expressions. The remainder of this Section examines a particular approach to evolving control rules in more detail.

SAMUEL is a machine learning program that uses a genetic algorithm and other competition-based heuristics to improve a set of control rules. The system explores alternative behaviours in simulation

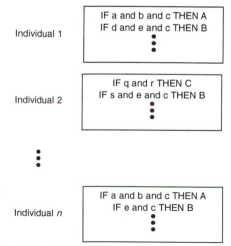

Figure 2.7   Evolving IF-THEN programs

and modifies its rules based on this experience. SAMUEL is designed for problems in which the payoff is delayed in the sense that payoff occurs only at the end of an episode which may span several decision steps. The primary features of SAMUEL are:

- A high-level rule language.
- Individual rules carry utility estimates based on experience; these estimates are used for conflict resolution.
- Genetic learning occurs at the level of rule sets, with specialised crossover and mutation operators.

The next Sections explain these features in more detail.

2.4.1 Knowledge representation

A distinctive feature of SAMUEL is that it employs a symbolic, attribute-value rule language, rather than the low-level representations adopted by many genetic algorithm-based systems. A typical control rule in SAMUEL is:

RULE R_i
       IF   time > 5
       AND   bearing = [ahead OR left]
       AND   1000 < range < 2000
       THEN set turn = 45   (*strength* 1.0)
       AND   set speed = high (*strength* 0.5)

Attributes can be either real-valued variables (e.g., time, range, turn) or nominal variables (e.g., bearing, speed). SAMUEL also supports fuzzy variables [5].

SAMUEL's rule language allows the user to specify an initial set of control rules which serves as a starting point for the genetic algorithm. The use of a symbolic rule language in SAMUEL is also intended to facilitate the incorporation of traditional machine learning methods into the system where appropriate. Experience with SAMUEL illustrates that genetic algorithms can be used to improve partially correct control rules, as well as to learn rules from scratch [21].

2.4.2 Rule strength

The strength in the rule estimates the utility of taking the specified action in situations matched by the rule's conditions. Rule strengths are modified through experience. When payoff is obtained, the strengths of all active rules (i.e., rules which suggested the actions taken during the current payoff period) are incrementally adjusted to reflect the current payoff. Over the course of many payoff episodes, the strengths of all active rules converge to the expected levels of payoff [11]. This observation motivates the use of rule strength as the deciding factor in conflict resolution. That is, if a number of rules suggest different actions at a given time, SAMUEL selects an action associated with a high strength rule.

2.4.3 Mutation operators

The high-level rule language used by SAMUEL provides an opportunity to alter rules using heuristic machine learning methods, in addition to random mutations. SAMUEL incorporates several symbolic rule modification operators, including:

- Specialisation: restricts the range in the conditions of a rule.
- Generalisation: expands the range in the conditions of a rule.
- Merge: combines two overlapping rules.

These mutation operators are Lamarckian in the sense that individual rules are modified based on the experience of the control strategy in the test environment. These changes are then passed along as 'genetic material' to subsequent generations of control strategies. The experience triggers for rule modification include:

- Specialisation: apply when low-strength general rules lead to good results. That is, the original rule appears to be over-generalised.
- Generalisation: apply when high-strength rules partially match. That is, the rule appears to be too specific.
- Merge: apply to two high-strength control rules which recommend the same action whose conditions overlap sufficiently.

Once created, a rule survives intact unless its rule set is not selected for reproduction or the rule is explicitly deleted by the deletion operator, which is applied to inactive, subsumed, or low-strength rules.

2.4.4 Crossover in SAMUEL

In SAMUEL, crossover forms new control rule sets by exchanging groups of rules between the selected parent rule sets. Rules that fire in sequence during a successful episode are inherited as a group, as shown in Figure 2.8. That is, crossover treats a successful rule sequence as a group during recombination. In this way, the offspring rule sets are likely to inherit some of the beneficial behaviour patterns of their parents. Of course, the success of the new combination of rules in the offspring depends on all of the other rules in the rule set.

Experience of Parent A:
Episode 1: $R_{A3} \rightarrow R_{A1} \rightarrow R_{A7} \rightarrow R_{A5}$ *High Payoff*
Episode 2: $R_{A2} \rightarrow R_{A6} \rightarrow R_{A8} \rightarrow R_{A4}$ *Low Payoff*

Experience of Parent B:
Episode 1: $R_{B7} \rightarrow R_{B1} \rightarrow R_{B5} \rightarrow R_{B5}$ *Low Payoff*
Episode 2: $R_{B6} \rightarrow R_{B2} \rightarrow R_{B4} \rightarrow R_{B4}$ *High Payoff*

Possible Offspring:
$\{ \ldots R_{A3}, R_{A1}, R_{A7}, R_{A5}, \ldots, R_{B6}, R_{B2}, R_{B4}, \ldots \}$

Figure 2.8   Crossover in SAMUEL

2.4.5 Control applications of SAMUEL

SAMUEL has been demonstrated on a variety of control tasks including evading a predator, stalking a prey, tracking moving targets and robotic navigation and collision avoidance [13].

In a recent study [22], SAMUEL was used to learn control rules for an autonomous mobile robot for a task in which the goal was to guide another mobile robot to a specified area. The control rules were learned under simulation, and the resulting rules were then used to control an operational mobile robot. Adding to the difficulty of this task, the control rules were required to avoid collisions with other objects in the world.

In this task, one robot played the role of a shepherd and the other robot represented a sheep. The sheep reacted to the presence of nearby objects by moving away from them. Otherwise, the sheep moved in a random walk. The shepherd's task was to control its own translation and steering to get the sheep to move into a pasture. Both robots were controlled by reactive rule sets which mapped current sensors into appropriate motion commands. Figure 2.9 shows an example of part of a SAMUEL rule set for the shepherd robot. Only the shepherd's rules were learned.

The performance of an individual rule set was calculated by averaging its relative success in the shepherding task over 20 episodes, in which each episode began with the sheep and the shepherd placed in random initial positions and orientations, and ended when the sheep entered the pasture, time expired, or a collision occurred. Over a period of 200 generations, SAMUEL improved the control rules for this task from an initial level of about 40 % successful to about 75 % successful.

RULE 1
     IF   front-sonar < 30
     AND   bearing > 10
     THEN set turn = 20   (*strength* 1.0)

RULE 2
     IF   front-ir < 5
     THEN set speed = -10   (*strength* 0.7)
  ...

Figure 2.9   Control rules for mobile robots

In order to verify the learned behaviours, the learned rules were used to control the actual shepherd robot. The shepherd robot and the sheep robot were placed in random locations and orientations within our laboratory environment, and the resulting performance recorded. In operational tests the shepherd robot succeeded in forcing the sheep robot to the desired location in 67 % of the episodes. Failures included the shepherd losing track of the sheep and communication failures between the computer and the robots, neither of which were accounted for in the simulation model. Removing these cases from consideration, the observed success rate was 73 %, very close to the success rate obtained under simulation.

Future work with SAMUEL will continue examining the process of building robotic systems through evolution. Scaling up will require better understanding of how multiple interacting behaviours can be evolved simultaneously. Other open problems include how to evolve hierarchies of skills and how to enable the robot to evolve new fitness functions as the need for new skills arises.

2.5 Evolutionary algorithms for testing intelligent control systems

The testing of control systems represents an important opportunity for evolutionary algorithms. Control systems for complex systems such as power plants or autonomous vehicles are likely to be based upon incomplete models of the plant and its operating environment. It may be difficult to predict the precise set of environmental conditions in which the plant will need to operate. Furthermore, intelligent control systems may include components such as expert systems which may include heuristic rules. As a result, testing and validation of such control systems may exceed the scope of current testing methods. However, if a simulation model of the plant is available, then an evolutionary algorithm can be used to help explore the range of conditions under which a control system will succeed or fail. One way of using an EA to test an intelligent control system is shown in Figure 2.10. The system incorporates four subsystems:

1 An evolutionary algorithm.
2 A model of the plant.
3 An intelligent control system for the plant.
4 A critic that evaluates the control system output.

The EA generates a population of test cases. The test cases may represent controllable parameters of the plant (e.g., control system setpoints, valve positions, pumps online), fault modes (e.g., pump failures, sensor failures) and environmental conditions (e.g., ambient air temperature). Each member of the population of test cases is submitted to the plant model. The plant model simulates the response of the plant to that input and generates a set of plant parameters, which represent the resulting state of the plant given the specified test cases.

The plant parameters are then presented to the intelligent control system, which diagnoses the cause of any performance problems in the plant and provides recommendations for appropriate control actions. The diagnoses and recommendations are submitted to the critic, which calculates a numerical value representing their quality. Based on a fitness function that is inversely related to the value returned by the critic, the EA generates additional test cases based on the fittest sets of current testing inputs and the cycle repeats. Since the fitness function used by the EA is inversely related to the quality of the expert system recommendation, the EA in effect searches for test cases which cause the expert system to behave poorly. That is, the EA is playing 'devil's advocate' to the intelligent control system.

This general method can be applied to a wide variety of control systems. Initial tests have included an automatic pilot for a flight simulator which controls a plane landing on an aircraft carrier, an intelligent controller for an experimental autonomous underwater vehicle [23] and an expert system for coal power plants [19]. Key issues for the successful application of this approach include:

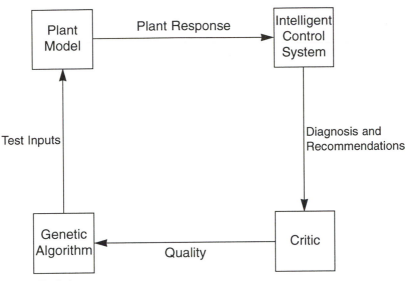

Figure 2.10   Evolving test cases

1   Development of the plant model. Ideally, the plant model should accurately predict the state of the plant for any test case presented to it by the evolutionary algorithm. Building a computational model for a complex plant presents many challenges, especially if it must predict plant behaviour for situations which are rarely seen during normal operation of the plant. To further complicate matters, the plant model must be computationally efficient, since the EA will require the evaluation of hundreds of test cases during the evolutionary process.

2   Specification of test cases. Ideally, the EA should only generate test cases which represent plausible situations for the plant. Unrealistic combinations of plant parameters, fault modes and environmental conditions should be avoided. On the other hand, the point of using an EA is that we want it to find unusual combinations of events which might escape consideration of other testing methods.

3   Design of the critic. The critic must evaluate the performance of the intelligent control system. Some aspects may be easy to measure (e.g., did the autopilot crash the plane?), but performance relative to an unknown ideal may be very difficult to measure (e.g., how well could the control system have performed, given this set of fault modes?).

Despite these challenges, evolutionary algorithms appear to be a promising addition to current methods for testing and evaluating complex systems.

2.6 Summary

The intent of this Chapter has been to outline some ways of applying evolutionary algorithms to control systems. It has been shown that EAs may interact with control systems at many levels, including the control parameter level, the levels of data structures such as schedules and routes, and the decision rule level. The SAMUEL system represents one approach to evolving control programs represented as IF-THEN rules. SAMUEL has advantages which include the ability to specify initial control rules as a starting point for the learning system and the ability to combine heuristic learning and genetic search. Finally, the use of evolutionary algorithms to search for challenging test cases for control systems was briefly described. Further developments along these lines can be expected to reduce the cost and effort required to build control systems with expert performance in complex domains.

2.7 Acknowledgment

This work is supported by the Office of Naval Research.

2.8 References

1   Baeck, T. and Schwefel, H.-P.: 'An overview of evolutionary algorithms for parameter optimisation', *Evolutionary Computation*, **1**, (1), pp 1-23
2   Bramlette, M., and Bouchard, E.: 'Genetic algorithms in parametric design of aircraft', in *Handbook of genetic algorithms*, Davis, L. (ed.), (Van Nostrand Reinhold: New York, 1991), pp 109-123
3   Chipperfield, A. and Fleming, P.: 'Gas turbine engine controller design using multiobjective genetic algorithms'. Proc. First IEE/IEEE international conference on *Genetic algorithms in engineering systems: innovations and applications* (GALESIA 95), Sheffield, UK, 1995, pp. 214–219
4   Chipperfield, A., and Fleming, P.: Genetic algorithms in control systems engineering', *J. of Computers and Control*, **24**, (1), 1996
5   Cobb, H. G., and Grefenstette, J. J.: 'Evolving fuzzy logic control strategies using SAMUEL: an initial implementation'. NCARAI technical report AIC-95-045, December 1995

6 Cox, L. A., Davis, L., and Qiu, Y.: 'Dynamic anticipatory routing in circuit-switched telecommunications networks', in *Handbook of genetic algorithms*, Davis, L. (ed.) (Van Nostrand Reinhold: New York, 1991), pp. 124-143

7 Dakev, N. V. and Chipperfield, A.: H-inf design of an EMS control system for a Maglev vehicle using evolutionary algorithms'. Proc. First IEE/IEEE international conference on *Genetic algorithms in engineering systems: innovations and applications* (GALESIA 95), Sheffield, UK, 1995, pp. 226–231

8 Davis, L. (ed.) *Handbook of Genetic Algorithms* (Van Nostrand Reinhold: New York, 1991)

9 De Jong, K. A.: 'Adaptive system design: a genetic approach', *IEEE Trans. Syst. Man Cybern.* 1980, **10**, (9), pp. 566-574

10 Grefenstette, J. J.: 'Optimisation of control parameters for genetic algorithms', *IEEE Trans Syst. Man Cybern.* 1986, **16**(1), pp. 122-128

11 Grefenstette, J. J.: Credit assignment in rule discovery system based on genetic algorithms. *Machine Learning* 1988, **3**(2/3), pp. 225-245

12 Grefenstette, J. J.: 'Genetic learning for adaptation in autonomous robots'. Proc. of the sixth international symposium on *Robotics and manufacturing* (ASME Press, New York, 1996)

13 Grefenstette, J. J., Ramsey, C. L. and Schultz, A. C.: 'Learning sequential decision rules using simulation models and competition', *Mach. Learn,* 1990, **5**, (4), pp. 355-381

14 Holland, J. H.: 'Escaping brittleness: the possibilities of general-purpose learning algorithms applied to parallel rule-based systems', in Michalski, R.S., Carbonell, J. G. and Mitchell, T. M. (eds.): *Machine learning: An artificial intelligence approach* (vol. 2) (Morgan Kaufmann, 1986)

15 Karr, C. L.: 'Design of an adaptive fuzzy logic controller using a genetic algorithm'. Proceedings of the fourth international conference on *Genetic algorithms*, San Mateo, CA, 1991, pp. 450-457, (Morgan Kaufmann)

16 Koza, J. R.: *Genetic programming* (MIT Press, 1992)

17 Kristinsson, K. and Dumont, G.: 'System identification and control using genetic algorithms'. *IEEE Trans. Syst. Man Cybern.*, 1992, **22**, (5), pp. 1033-1046

18 Ram, A., Arkin, R., Boone, G. and Pearce, M.: 'Using genetic algorithms to learn reactive control parameters for autonomous robotic navigation', *Adapt. Behav.* 1994, **2**, (3), pp. 277-305

19 Roache, E., Hickok, K., Loje, K., Hunt, M. and Grefenstette, J.: Genetic algorithms for expert system validation, Proceedings of the

1995 Western Multiconference Society for *Computer Simulation*, Las Vegas, NE, January 1995

20 Salami, M., and Cain, G.: 'An adaptive PID controller based on genetic algorithm processor'. Proc. first IEE/IEEE international conference on *Genetic algorithms in engineering systems: innovations and applications* (GALESIA 95), Sheffield, UK, 1995, pp. 88-93

21 Schultz, A. C. and Grefenstette, J. J.: 'Improving tactical plans with genetic algorithms'. Proceedings IEEE conference on *Tools for AI 90*, Washington, DC, 1990, pp. 328-334

22 Schultz, A. C., Grefenstette, J. J. and Adams, W.: 'Learning complex robotic behaviors'. Proc. of the sixth international symposium on *Robotics and manufacturing* (ASME Press, New York, 1996)

23 Schultz, A. C., Grefenstette, J. J. and De Jong, K. A.: 'Test and evaluation by genetic algorithms', *IEEE Expert*, **8**, (5), pp. 9-14, 1993

24 Smith, F. F.: 'Flexible learning of problem solving heuristics through adaptive search'. Proceedings of the eighth international joint conference on *Artificial intelligence*, Karlsruhe, Germany 1983, pp. 422-425 (Morgan Kaufmann)

25 Tan, K., Li, Y., Murray-Smith, D. and Sharman, K.: 'System identification and linearisation using genetic algorithms with simulated annealing'. Proc first IEE/IEEE international conference on *Genetic algorithms in engineering systems: innovations and applications* (GALESIA 95), Sheffield, UK, 1995, pp. 164-169

26 Varsek, A., Urbacic, T. and Filipic, B.: 'Genetic algorithms in controller design and tuning', *IEEE Trans. Syst. Man Cybern.*, 1993, **23**, (5), pp. 1330-1339

27 Yamada, T., and Nakano, R. 'A genetic algorithm with multi-step crossover for job-shop scheduling problems'. Proc. first IEE/IEEE international conference on *Genetic algorithms in engineering systems: innovations and applications* (GALESIA 95), Sheffield, UK, pp. 146-151

28 Yamauchi, B., and Beer, R.: 'Sequential behavior and learning in evolved dynamical neural networks,' *Adapt. Behav.*, 1994, **2**, (3), pp. 219-246

Chapter 3
Multiobjective genetic algorithms

C. M. Fonseca and P. J. Fleming

The populations of current approximations maintained by genetic algorithms (GAs) and other evolutionary approaches confer the ability to concurrently search for multiple solutions to given problems. This is particularly relevant in engineering, where multiple and often conflicting objectives seldom define optimal solutions uniquely. However, this ability is overlooked in most current applications of GAs in engineering, and GAs are used simply for their generality and robustness as an alternative to, but in the same spirit of, more restrictive conventional optimisers. Different objectives are thus analytically combined into a single function prior to optimisation, and the GA applied.

This Chapter aims to illustrate how an existing GA can be modified and set up to explore the relevant trade-offs between multiple objectives with a minimum of effort. Although Pareto and Pareto-like ranking schemes [1, 2] can be easily implemented, current guidelines on the associated set up of techniques such as sharing and mating restriction [3, 2] are intricate and/or based on more or less rough assumptions about the cost landscape, which has not contributed to their popularity.

However, if fitness sharing is reinterpreted as a technique involving the estimation of the population density at the points defined by each individual by so-called kernel methods [4], the setting of the sharing parameter comes to depend only on the size and current distribution of the population, and not on the problem. Kernel density estimation [4], a technique from statistics and data analysis, will be introduced and shown to find direct application in sharing and mating restriction, simplifying implementation and avoiding the introduction of any more tunable parameters in the GA formulation.

After a brief introduction to multiobjective optimisation and a discussion of preference articulation in GAs, the main differences between single-objective and multiobjective GAs are highlighted, and the conversion of an existing GA into a multiobjective GA is described by means of an example. Simple experimental results are presented towards the end of the Chapter for the purpose of illustration.

3.1 Multiobjective optimisation and preference articulation

Most engineering problems are characterised by several noncommensurable and often competing objectives to be optimised. Due to the trade-offs involved, such problems usually have no unique, perfect solution. Instead, they admit a set of equally valid, or nondominated, alternative solutions, which is known as the Pareto-optimal set [5]. These solutions are such that improvement in any objective can only be achieved at the expense of degradation in other objectives, and can only be discriminated on the basis of expert knowledge of the problem. This may include the understanding of the importance of certain objectives relative to others or the need to meet given specifications, for example.

Although nondominated solutions can generally be obtained through optimisation, expressing informal design preferences in terms of a sufficiently well behaved cost function, as expected by many conventional optimisers, is not always easy. In particular, unimodality requirements imply that all decisions must be made prior to optimisation. If the solution produced by the optimizer is not satisfactory, the cost function must be changed and the process repeated.

On the other hand, genetic algorithms only require that the cost (or, alternatively, the utility) of each individual be determined with respect to the current population so as to permit the broad ranking of the population. Individuals need only be rated better than, similar to, or worse than others, effectively allowing the decision maker to delay otherwise uninformed decisions until sufficient insight into the problem has been gained. At that point, the decision maker can adjust the current decision strategy, as the population evolves.

3.2 How do MOGAs differ from simple GAs?

In single-objective GAs, individual performance, as measured by the objective function, and individual fitness are so closely related that the objective function is sometimes referred to as the fitness function. The two are, however, not the same. In fact, whereas the objective function characterises the problem and cannot be changed at will, assigned fitness is a direct measure of individual reproductive ability (i.e., expected number of offspring), forming an integral part of the GA search strategy.

This distinction becomes all the more important when performance is measured in terms of a vector of objective values, because fitness must remain a scalar. In this case, fitness assignment is a more elaborate process. For the sake of generality, the necessary scalarisation of the objective vectors may be viewed as a multicriterion decision making problem involving a (finite) number of candidates, the individuals in the population [2, 6]. Individuals are thus assigned a measure of their utility indicating whether they perform better, worse or similarly to others, and possibly also how much better or worse. If the utility measure conveys only ordinal information, then fitness must be assigned through ranking. Otherwise, ranking or proportional fitness assignment may be used. This setup is general enough to include problems where individual performance must be assessed through pairwise comparison [7], such as when evolving game-playing programs.

Since the solution of a multicriterion decision making problem depends only on the vectorial performance of the available candidates and on the preferences of the decision maker, and not on any subsequent search or optimisation, utility is also essentially different from fitness. In particular, techniques such as sharing affect the individuals' fitness, but not their utility or cost.

3.2.1 Scale-independent decision strategies

In the total absence of information concerning the relative importance of the objectives, Pareto dominance is the only basis on which an individual can be said to perform better than another. Therefore, nondominated individuals must all be considered best performers, and thus be assigned the same cost [1], e.g., zero. Deciding on the performance of dominated individuals is a more subjective matter. One may, for example, assign to them a cost proportional to how many individuals in the population dominate them (Figure 3.1), in which case nondominated individuals would also be treated as desired. This is essentially the Pareto ranking scheme proposed in [2].

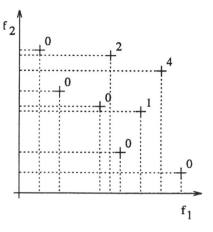

Figure 3.1   Pareto ranking

Another popular Pareto ranking scheme [1], also known as non-dominated sorting [8], consists of removing the nondominated individuals (here also ranked zero, for ease of comparison) from contention, finding the nondominated individuals in the remaining population and assigning them rank 1, and so forth, until the whole population is ranked. Both of these approaches guarantee that non-dominated individuals are all ranked best, and that individuals are consistently assigned better ranks than those which they dominate. However, the first ranking scheme does appear to be easier to interpret and analyse mathematically [6].

When goal and/or priority information is available for the objectives, it may become possible to discriminate between some non-dominated solutions. For example, if degradation in objective components which meet their goals does not go beyond the goal boundaries, and results in the improvement of objective components which do not yet satisfy the corresponding goals, then it should be accepted. Similarly, in a dual priority setup [6], it is only important to improve on high priority objectives (i.e., constraints) until the corresponding goals are met, after which improvement should be sought for the remaining objectives. These considerations have been formalised [6] in terms of a transitive relational operator (preferability), based on Pareto dominance, but which selectively excludes objectives according to their priority and to whether or not they meet their goals.

For simplicity, only one level of priority will be considered here. The full, multiple priority version of the preferability operator is described in detail in [6]. Consider two objective vectors $\vec{u}$ and $\vec{v}$ and a goal vector $\vec{g}$. Also, let $\breve{u}$ (u-smile) and $\hat{u}$ (u-frown) denote the components of $\vec{u}$ which meet their goals and those which do not, respectively. Assuming minimisation, one can write:

$$\vec{u}^{\breve{u}} \leq \vec{g}^{\breve{u}} \wedge \vec{u}^{\hat{u}} > \vec{g}^{\hat{u}}$$

where the inequalities apply componentwise. This is equivalent to:

$$\forall i \in \breve{u}, u_i \leq g_i \wedge \forall i \in \hat{u}, u_i > g_i$$

where u_i and g_i represent the components of $\vec{u}$ and $\vec{g}$, respectively. Then, $\vec{u}$ is said to be preferable to $\vec{v}$ given $\vec{g}$ if and only if:

$$\left(\vec{u}^{\hat{u}} p{<} \vec{v}^{\hat{u}} \right) \wedge \left\{ \left(\vec{u}^{\hat{u}} {=} \vec{v}^{\hat{u}} \right) \wedge \left[\left(\vec{v}^{\breve{u}} {\leq} \vec{u}^{\breve{u}} \right) \vee \left(\vec{u}^{\breve{u}} p{<} \vec{v}^{\breve{u}} \right) \right] \right\}$$

where $\vec{a} p{<} \vec{b}$ denotes $\vec{a}$ dominates $\vec{b}$. In other words, $\vec{u}$ will be preferable to $\vec{v}$ if and only if one of the following is true:

(i) The violating components of $\vec{u}$ dominate the corresponding components of $\vec{v}$.

(ii) The violating components of $\vec{u}$ are equal to the corresponding components of $\vec{v}$, but $\vec{v}$ violates at least another goal.

(iii) The violating components of $\vec{u}$ are equal to the corresponding components of $\vec{v}$, but $\vec{u}$ dominates $\vec{v}$ as a whole.

Like Pareto dominance, this relation can be used to rank the individuals in a population by one of the methods described above.

3.2.2 Cost to fitness mapping and selection

Once cost has been assigned, selection can proceed in much the usual way. Suitable alternatives include rank-based cost to fitness mapping [9] followed by stochastic universal sampling [10] (or even roulette wheel selection) and tournament selection also based on cost, as reported in [11, 12].

Exponential rank-based fitness assignment is illustrated in Figure 3.2. Individuals are sorted by cost (the values are those from Figure 3.1) and first assigned fitness values according to an exponential rule (narrower bars). Then, a single value of fitness is derived for each group of individuals with the same cost, through averaging (wider bars).

3.2.3 Sharing

Although all preferred individuals are assigned the same fitness, their actual number of offspring, which must obviously be an integer, may

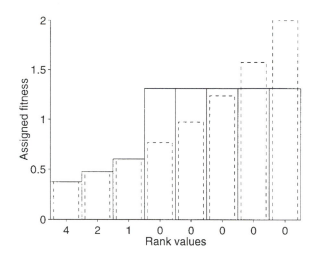

Figure 3.2   Rank-based fitness assignment

differ. The imbalance can easily accumulate with the generations and result in the population drifting towards an arbitrary region of the trade-off surface, a phenomenon known as genetic drift [13]. In addition, recombination and mutation may be less likely to produce individuals in certain regions of the trade-off surface (e.g., the extremes) than in others, causing the population to cover only a small part of it.

Fitness sharing [14], originally introduced to promote the sampling of multiple fitness peaks, helps to counteract genetic drift by penalising individuals due to the presence of other individuals in their neighbourhood. The niche count of each individual is initially set to zero and then incremented by a certain amount for every individual in the population, including itself. The contribution of an individual to another's niche count is dictated by a sharing function, which is a function of their mutual distance in genotypic, phenotypic or objective space. Raw fitness values are then weighted by the inverse of the niche counts and subsequently normalised by the sum of the weights, before selection. In this way, the total fitness in the population is redistributed (and thus shared) by the individuals. Fitness can also be shared only between individuals with the same raw fitness, by computing partial weight totals and performing the normalisation within each group of such individuals [2].

The use of fitness sharing has been restricted by the difficulty found in determining the appropriate niche size, i.e., how close together individuals should be for degradation to occur. Current guidelines either make assumptions about the number and distribution of peaks in the cost landscape [3], or rely on the estimation of the (maximum) size of the trade-off surface based on the properties of the Pareto set [2].

However, niche count computation (explained above) turns out to be remarkably similar to the kernel density estimation methods [4] known to statisticians. Basically, density estimates are computed in exactly the same way as niche counts, except for a constant factor. The parallel is drawn in Table 3.1.

Table 3.1 The analogy between sharing and kernel estimation

Fitness sharing	Kernel density estimation
sharing function	kernel function
niche size (σ_{share})	smoothing parameter (h)
niche count	density estimate

As in sharing, the choice of the smoothing parameter is ultimately subjective, but guidelines have been developed for certain kernels, such as the standard normal probability density function and the Epanechnikov kernel. The latter can be written as [4]:

$$K_e(d/h) = \begin{cases} \dfrac{1}{2} c_n^{-1}(n+2)\left[1+(d/h)^2\right] & \text{if } d/h < 1 \\ 0 & \text{otherwise} \end{cases}$$

where n is the number of decision variables, c_n is the volume of the unit n-dimensional sphere and d/h is the normalised Euclidean distance between individuals. The parameter h is the smoothing parameter analogous to σ_{share}. Note that this kernel is, apart from the constant $c_n^{-1}(n+2)/2$, a particular case of the family of power law sharing functions proposed by Goldberg and Richardson [14].

According to Silverman [4], a good choice (approximately optimal in the least mean integrated squared error sense if the population follows a multivariate normal distribution) of the smoothing parameter for the Epanechnikov kernel $K_e(d)$ is:

$$h = \left[8 c_n^{-1}(n+4)(2\sqrt{\pi})^n / N\right]^{1/(n+4)}$$

for a population with N individuals and identity covariance matrix. Populations with arbitrary sample covariance matrix S can simply be sphered (or normalised) by multiplying each individual by a matrix R such that $RR^T = S^{-1}$. This implies that the niche size (which depends on *h* and *S*) can be constantly, and automatically, adapted to suit the population at each generation, regardless of what the cost function may be.

These results can be used directly to perform sharing in Euclidean decision variable spaces. It might be possible to develop guidelines based on the same principles for other types of spaces.

3.2.4 Mating restriction

Mating restriction consists of biasing the way in which individuals are paired for recombination [3]. As the population distributes itself along the trade-off surface, recombining arbitrary pairs of individuals may be conducive to the formation of a large number of unfit offspring, or lethals. To address this issue, mating can be restricted, where possible, to individuals within a given distance σ_{mate} from each other. Following the common practice of setting $\sigma_{mate} = \sigma_{share}$, individuals may be allowed to mate only if they lie within a distance *h* from each other in the sphered space used for sharing.

3.2.5 Interactive optimisation and changing environments

As the GA population evolves and trade-off information is acquired, the decision maker may wish to see the population concentrate on a smaller region of the trade-off surface, or even back off and move on to a totally different region. This can be achieved simply by changing the goals supplied to the GA at the cost assignment stage, which in turn affects the ranking of the population and modifies the cost landscape. The GA must then be able to respond quickly to such preference changes.

Introducing a small percentage (10–20 %) of random individuals at each generation has been shown to make the GA more responsive to sudden changes in the fitness landscape [15]. This technique can be easily incorporated in a multiobjective GA.

3.3 Putting it all together

The implementation of a multiobjective GA incorporating the techniques described in the previous Section will now be considered. Matlab [16] pseudocode for a simple, aggregating, GA is given in Figure 3.3. Individual chromosomes (the rows of matrix Chrom) are initially generated at random, and then decoded, producing the corresponding vectors of decision variables, in matrix DVar. Evaluation is made in two steps: objective vectors are computed first (rows of ObjV), and then aggregated to produce a scalar measure of cost for each individual (the components of vector Cost). Fitness is assigned through ranking, with given selective pressure, SP. Individuals are selected using SUS (stochastic universal sampling), recombined and mutated, and a new generation begins. Functions multobjfun and aggregate are written by the user, the former defining the problem and the latter implementing a fixed decision strategy, such as a weighted sum. The remaining functions implement the GA itself, and are not far from those found in the current version of the GA Toolbox for Matlab [17].

```
Chrom = creatpop(NIND, LIND);
while Gen < MAXGEN
        DVar = decode(Chrom);
        ObjV = multobjfun(DVar);
        Cost = aggregate(ObjV);
        Fitn = ranking(Cost, SP);
        Ix = sus(Fitn);
        SelCh = Chrom(Ix, :);
        SelCh = xover(SelCh, XOVR);
        Chrom = mutate(SelCh, MUTR);
        Gen = Gen + 1;
end
```

Figure 3.3   A simple aggregating GA

Preference-based multiobjective ranking (rank_prf in Figure 3.4) comes as a drop-in replacement for aggregate which may take two optional parameters: a goal vector, GoalV, and a vector indicating the priority of the objectives, PriorV (not used in the example).

Niche counts NicheC are computed using a kernel estimator based on the Epanechnikov kernel. DVar is passed to the function twice because it constitutes simultaneously the sample data and the points where the population density needs to be estimated. The estimation function also returns the default smoothing parameter Sigma (h) and a matrix R such that DVar*R has identity covariance matrix, both of which are used at a later stage in mating restriction. The ranking function now uses NicheC to perform sharing between individuals with equal cost as an integral part of the fitness assignment procedure.

```
Chrom = creatpop(NIND, LIND);
while Gen < MAXGEN
    DVar = decode(Chrom);
    ObjV = multobjfun(DVar);
    Cost = rank_prf(ObjV, GoalV);
    [NicheC, Sigma, R] = epanechnikov(DVar, DVar);
    Fitn = ranking(Cost, SP, NicheC);
    Ix = sus(Fitn, NIND – NImmigr);
    SelCh = Chrom(Ix, :);
    SelDV = DVar(Ix, :);
    PermIx = pairup(SelDV * R, Sigma);
    SelCh = SelCh(PermIx, :);
    SelCh = xover(SelCh, XOVR);
    Chrom = [ mutate(SelCh, MUTR);
                creatpop(NImmigr, LIND) ];
    Gen = Gen + 1;
end
```

Figure 3.4   A multiobjective GA

Since a small number NImmigr of individuals in the new population will consist of random immigrants, only NIND-NImmigr individuals are selected from the old population. Mating restriction is implemented by reordering the individuals in the population so that consecutive pairs of chromosomes in SelCh correspond, where possible, to individuals within a required distance Sigma of each other in normalised decision variable space. (The parental population is rotated and scaled according to the same transformation, R, used for niche count computation.) The random immigrants are appended to the population after mutation, having to survive selection before being allowed to recombine. This will be most likely whenever the fitness landscape changes and the GA population is no longer adapted to it.

As can be easily seen, the only additional GA parameter in this second version of the GA is the number of random immigrants to be inserted in the population at each generation, the setting of which is not critical. Random immigrants make the GA more exploratory and thus more responsive to sudden preference changes, as long as a balanced amount of exploitation can still be maintained. In particular, selective pressure should probably be increased slightly, to compensate for the fact that the insertion of random immigrants into the population reduces the expected number of offspring of the best individual by NImmigr/NIND.

3.4 Experimental results

Several applications of multiobjective GAs have been reported in the literature, mainly related to control engineering. In an independent study, Whidborne *et al.* [18] have compared a multiobjective GA based on the preferability relation to other interactive multiobjective approaches such as the method of inequalities, and noted the tendency for the MOGA to produce solutions very similar to each other. However, they also pointed out that the GA did not include sharing or mating restriction.

To show how sharing and mating restrictions together can significantly contribute to the performance of the GA, consider the minimisation of the following two objectives:

$$f_1(x_1,...x_n)=1 - \exp\left[-\sum_{i=1}^{n}\left(x_i - 1/\sqrt{n}\right)^2\right]$$

$$f_2(x_1,...x_n)=1 - \exp\left[-\sum_{i=1}^{n}\left(x_i + 1/\sqrt{n}\right)^2\right]$$

which are defined for any number of decision variables n. The minimum of f_1 is located at $(x_1,...x_n)=(1/\sqrt{n},...,1/\sqrt{n})$ for all n, and that of f_2 is located at $(x_1,...x_n)=(-1/\sqrt{n},...,-1/\sqrt{n})$. Due to the symmetry of the two functions, the Pareto-optimal set clearly corresponds to all points on the line defined by:

$$x_1=x_2=\cdots=x_n \wedge -1/\sqrt{n}\leq x_i\leq1/\sqrt{n}$$

A simple genetic algorithm with a population size of 100 individuals, binary chromosomes, reduced-surrogate shuffle crossover and binary mutation was used to approach this problem for $n=8$. Decision variables were Gray-encoded as 16-bit strings in the interval [−2,2] and concatenated to form the chromosomes. Multiobjective ranking was performed as described and illustrated earlier in Figure 3.1.

Running this GA for 100 generations, without sharing or mating restriction, shows how the population tends to concentrate on a small region of the trade-off surface (Figure 3.5). Nondominated individuals are marked with filled circles (●) and other individuals with empty circles (o). The solid line represents the best approximation to the real trade-off surface (dashed line) known as a consequence of the GA run.

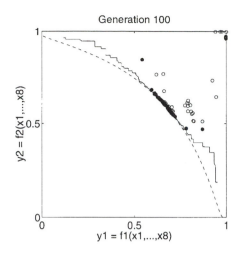

● Nondominated individuals
○ Dominated individuals
—Best trade off found (cumulative)
— — Actual Pareto set

Figure 3.5   Multiobjective GA without sharing or mating restriction

If, however, sharing and mating restrictions are implemented in the decision variable domain as in the example in Figure 3.4, the population is able to remain distributed across the whole trade-off surface. This can be seen in Figure 3.6.

The population can also be driven to sample a given region of the trade-off surface by setting goals accordingly. Figure 3.7 shows the distribution of the population after setting the goals and running the GA for another 50 generations. Most of the population can be seen to be concentrated in the preferred region of the trade-off surface, as desired.

An application of the multiobjective GA described here in the design of controllers for gas turbine engines is reported in [19].

3.5 Concluding remarks

Although multiobjective genetic algorithms are still undergoing development, their application to real world problems is becoming increasingly feasible. In combination with a suitable graphical user interface, multiobjective GAs can become a powerful, and possibly

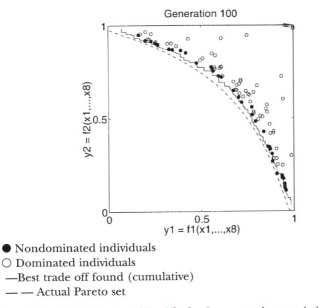

Figure 3.6 *Multiobjective GA with sharing or mating restriction*

interactive, decision support tool, allowing a decision maker to learn about the problem before committing to a final decision.

Due to the multisolution nature of most multiobjective problems, fitness sharing is needed to maintain diversity in the population. However, guidelines on how to set the sharing parameter have been too dependent on suppositions about the fitness landscape which are difficult to make in any practical setting. Understanding sharing as something similar to density estimation can make the use of sharing, and thus that of multiobjective GAs, more practical. More elaborate density estimation techniques, such as adaptive kernel density estimation [4], may further improve the quality of sharing. On the other hand, nearest-neighbour estimators may be easier to extend to nonEuclidean spaces, and thus be more appropriate to ordering and grouping problems, for example.

Finally, multiobjective evolutionary optimisation is a much broader area than reported here, and the interested reader is referred to [20] for an overview.

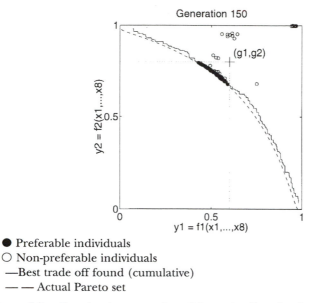

Generation 150

y2 = f2(x1,...,x8)

y1 = f1(x1,...,x8)

(g1,g2)

● Preferable individuals
○ Non-preferable individuals
—Best trade off found (cumulative)
— — Actual Pareto set

Figure 3.7   Zooming in on a region of the trade-off surface by setting goals accordingly

3.6 Acknowledgment

The authors wish to acknowledge the support of the UK Engineering and Physical Sciences Research Council (grant GR/J70857).

3.7 References

1   Goldberg, D. E.: *Genetic algorithms in search, optimisation and machine learning* (Addison-Wesley, 1989)
2   Fonseca, C. M., and Fleming, P. J.: 'Genetic algorithms for multi-objective optimisation: Formulation, discussion and generalisation,' in *Genetic algorithms: proceedings of the fifth international conference,* Forrest, S. (ed.) (Morgan Kaufmann, 1993), pp. 416–423
3   Deb, K., and Goldberg, D. E.: 'An investigation of niche and species formation in genetic function optimisation,' in *Proceedings of the third international conference on genetic algorithms* Schaffer, J. D. (ed.), (Morgan Kaufmann, 1989), pp. 42–50.
4   Silverman, B. W.: Density estimation for wtatistics and eata analysis, vol. 26 of *Monographs on statistics and applied probability.* (Chapman and Hall, 1986)

5   Ben-Tal, A.: 'Characterisation of Pareto and lexicographic optimal solutions,' in *Multiple criteria decision making theory and application* Fandel, G. and Gal, T. (eds.), vol. 177 of *Lecture Notes in Economics and Mathematical Systems*, (Springer-Verlag, 1980), pp. 1–11

6   Fonseca, C. M., *Multiobjective genetic algorithms with application to control engineering problems*, Ph.D. Thesis, Dept. Automatic Control and Systems Eng., University of Sheffield, Sheffield, U.K., 1995

7   Uppuluri, V. R. R., 'Prioritisation techniques based on stochastic paired comparisons,' in *Multiple criteria decision making and risk analysis using microcomputers*, Karpak, B. and Zionts, S., (eds.), vol. 56 of *NATO ASI Series F: Computer and Systems Sciences*, (Springer-Verlag, 1989), pp. 293–303

8   Srinivas, N. and Deb, K., 'Multiobjective optimisation using nondominated sorting in genetic algorithms,' *Evolutionary Computation*, vol.2, 1994

9   Baker, J. E.: 'Adaptive selection methods for genetic algorithms,' in *Genetic algorithms and their applications: proceedings of the first international conference on genetic algorithms* Grefenstette, J. J., (ed.) (Lawrence Erlbaum, 1985) pp. 101-111

10 Baker, J. E.: 'Reducing bias and inefficiency in the selection algorithm,' in [21], pp. 14-21.

11 Cieniawski, S. E.: 'An investigation of the ability of genetic algorithms to generate the tradeoff curve of a multi-objective groundwater monitoring problem,' Master's thesis, University of Illinois at Urbana-Champaign, Urbana, Illinois, 1993

12 Ritzel, B. J., Eheart, J. W. and Ranjithan, S.: 'Using genetic algorithms to solve a multiple objective groundwater pollution containment problem,' *Water Resour Res*, **30**, pp. 1589-1603, 1994

13 Goldberg, D. E. and Segrest, P.: 'Finite markov chain analysis of genetic algorithms,' in [21], pp.1-8.

14 Goldberg, D. E. and Richardson, J.: 'Genetic algorithms with sharing for multimodal function optimisation,' in [21], pp. 41-49

15 Grefenstette, J. J.: 'Genetic algorithms for changing environments,' in *Parallel problem solving from nature*, 2, Maenner, R. and Manderick, B. (eds.), (North-Holland, 1992), pp. 137-144

16 The MathWorks, Inc., *Matlab Reference Guide*, August 1992

17 Chipperfield, A., Fleming, P., Pohlheim, H. and Fonseca, C., 'Genetic algorithm toolbox user's guide,' Research report 512, Dept. Automatic Control and Systems Eng., University of Sheffield, Sheffield, U.K., July 1994.

18 Whidborne, J. F., Gu, D.-W. and Postlethwaite, I.: 'Algorithms for the

method of inequalities – a comparative study,' in Proc. American *Control* Conference, Seattle, USA, 1995

19 Chipperfield, A. J. and Fleming, P. J. 'Multiobjective gas turbine engine controller design using genetic algorithms,' *IEEE Trans. Ind. Electron,* **45** (5) pp. 583-587, 1996

20 Fonseca, C. M. and Fleming, P. J. 'An overview of evolutionary algorithms in multiobjective optimisation,' *Evolutionary Computation,* **3**, pp. 1-16, Spring 1995

21 Grefenstette, J. J. (ed.): *Genetic algorithms and their applications: proceedings of the second international conference on genetic algorithms* (Lawrence Erlbaum, 1987)

Chapter 4
Constraint resolution in genetic algorithms
R. Pearce

4.1 Introduction

Most real world optimisation problems have constraints. Solutions that appear to be very good on the basis of the objective criteria, may be unacceptable for some other reason. For example, a design optimised to minimise cost must also meet stress and manufacturing requirements. If the cheapest design buckles under load or is impossible to make, then it is not an acceptable solution. What is actually required is the cheapest design that meets these additional criteria.

The problem of including these constraints into a search is common to all optimisation techniques. In linear problems, optimal solutions may be found by following the constraint boundaries. Constraint satisfaction methods are used to find solutions in highly constrained problems, although they are impractical for finding the optimum if the search space is large.

It is clear, however, that other methods are required for applying constraints to the large, ill-behaved spaces typically searched by a genetic algorithm. These constraints are important in solving applications, whether design, scheduling, system identification or control, or any of the myriad of areas to which genetic algorithms have been applied.

4.2 Constraint resolution in genetic algorithms

When an optimisation problem is subject to constraints, then these must be incorporated into the search algorithm. There are a number of ways to ensure that the constraints are taken into account during the optimisation. It may be possible to restrict the search to valid regions of the search space. The maximum and minimum of each parameter in the chromosome string are set to sensible values in the light of knowledge about the system. In addition, the chromosome may be encoded or manipulated in a way which prevents the generation of invalid solutions. Unable to generate solutions that do not meet the constraint criteria, the algorithm avoids wasting effort evaluating solutions which would not be acceptable.

However, in many cases, some calculation is required to determine whether constraints have been met. Examples of this could be the power output of a generator or the stresses within a component. As these values are a function of the solution as a whole, there is no way of preventing the optimiser generating solutions which violate these constraints. Therefore, the search is directed to valid regions by penalising the fitness of solutions that violate the constraints [3]. Setting the fitness of any invalid solution to zero ensures that only valid solutions are considered for breeding and that the final solution will meet the constraints. However, this can result in valuable information being lost from the gene pool as invalid solutions may still contain fit schema.

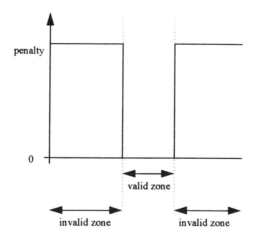

Figure 4.1   Fixed penalty for constraint violation

It is more usual to reduce the fitness by a finite amount. The lower fitness reduces the solution's chance of breeding with respect to those that do meet the constraints. However, the genes still have a chance of being passed on to successive generations. The method of using a fixed penalty (see Figure 4.1) for any degree of violation is rarely applied as it provides no ranking of solutions which violate the constraint. This is because a solution which has only just failed to meet a constraint is penalised to the same extent as one that has missed by a long way. This is less important if the constraints are easily met. However, if it is hard to find a solution that meets the constraints, then the search will flounder around the search space until, by chance, it lands in a valid region.

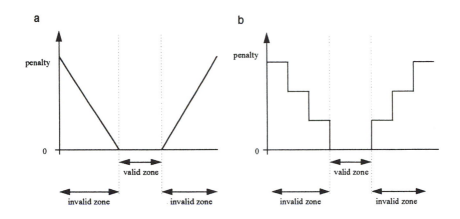

Figure 4.2   *Variable penalty for constraint violation*
          a linear variation of penalty with constraint violation
          b stepped penalty with constraint violation

It is more common to use a graduated scale of penalty, proportional to the magnitude of the constraint violation. The form of the penalty will depend on the problem and the importance of the constraint. For example, it may be linear (see Figure 4.2(a)), polynomial, exponential or stepped (see Figure 4.2(b)) [4]. The stepped approach will tend to give the same rank to several solutions (like the single step function of Figure 4.1) unless the step width is small with respect to the discretisation of the objective. The exponential form is closely equivalent to simulated annealing [5], where solutions which violate constraints may be selected with a probability that varies exponentially with the size of the violation. In all cases, as the search moves towards solutions that meet the constraints, the penalties reduce. Solutions that only just fail to meet the constraints have an increased chance of being used as parents. Hence, the search will move towards valid regions in the search space.

4.3 Problems in encoding of constraints

There is a significant degree of tuning required to balance the effects of the constraints and the objective function. The magnitude of the penalty and its variation with the degree of violation have to be adjusted to give the required response over the entire search space. Furthermore, it may be appropriate to adjust the magnitude of the penalty during the course of the optimisation. A small penalty widens the size of the target volume early during the search, and a large penalty late in the search forces closer adherence to all the constraints. When setting up the objective function and constraints it is important to remember that a genetic algorithm will only seek to maximise its fitness. If the effect of a constraint violation is to increase the fitness to a level that compensates for the incurred penalty, then this will occur. In this way, trade off can occur between constraints, or between the objective value and constraints. In some cases, this may be desirable, in others not so. The difficulty is in controlling the objective function to give the required characteristics.

Also important is the relative weighting of the constraints. It may be more critical to meet some constraints than others. Some constraints (hard constraints) may need to be met at any cost, regardless of their effect on fitness or other constraints. Other constraints (soft constraints) may be more flexible, acting like a wish list of attributes to which the user would like the solutions to conform. The magnitude and importance of these constraints may be either independent of other constraints and objectives or interrelated. In other words, a minor violation on one soft constraint may be acceptable if it gives a significant benefit in terms of fitness, or if other constraints have been comfortably met. This form of reasoning is difficult to hard encode into the objective function. It is also difficult to verify once encoded, as the logic is hidden. This is illustrated in the following equation:

$$\text{fitness} = \text{objective} + \sum \frac{cons_i}{datum_i} + (cons_1 * cons_2)$$

Even if a valid balance between the objective and constraint values is obtained, there may be a number of residual problems.

There may be no solution that can satisfy all the constraints. In this case, the constraints become objectives in themselves, with the genetic algorithm attempting to minimise the violations. Under these circumstances, the relative weighting of the constraints and the degree to which constraint achievements affect the acceptability of other constraint violations becomes particularly important. For example, if it is not possible to meet all the stress and thermal constraints in the design of a component, then stress constraints may be relaxed in regions of low temperatures. A solution that has just failed in one constraint may be considered as being worse than one that has just failed to meet two other constraints if the first constraint is a critical one. The degree to which a violation is acceptable may depend on how many or which other constraints have been met, or the degree to which it benefits the objective value. For example, a slightly higher stress in a critical region of a component may be more significant than if it was in a less critical region.

There may be insufficient data and hard constraints to identify a unique solution. However, qualitative guides may be available to direct the search towards the desired region. This frequently exists in the form of engineering judgement. This is the knowledge used by people experienced in the field that directs their path during a manual optimisation. This could be something like 'I would expect those two parameters to be about equal', or that 'the length of one arm of the component will probably be about twice the length of the other arm'.

The problem may be ill conditioned and so very sensitive to noise [8]. This can make it difficult to solve accurately analytically. There may be a few qualitative constraints available to reduce the effect of the ill-conditioned nature.

In all the cases, mathematical descriptions of the constraints can be difficult to formulate and more difficult to audit.

4.4 Fuzzy encoding of constraints

During manual optimisation, knowledge and engineering judgement are applied to determine the overall quality of the solution. This is rarely done in absolute numerical terms, but rather in a qualitative manner. The engineer takes into account all aspects of the design to decide whether a solution is better than another. The engineer may be prepared to compromise his requirements in some aspects when the ideal is unattainable. These compromises are usually formed by approximately classifying the various aspects in respect to the requirements. Hence, a design that has greatly exceeded the pressure requirements might be allowed a very small violation of a temperature requirement. These descriptions will correspond to approximate ranges of values, but there are no hard cut offs. A small violation does not become a large violation simply because it has increased from 0·9 to 1·1 if the boundary between the two has been set to 1·0. We think of the classifications in fuzzy terms, because that is the way that our brains work.

The use of these qualitative judgements to resolve objectives and constraints has the potential for resolving many of the areas of concern given in Section 4.3. Initial studies have been performed on a number of artificial toy problems as well as a genuine real world application in industry. These are described in later Sections (Sections 4.6 and 4.7). In order to facilitate this understanding, the following section gives a brief, and nonmathematical, overview of fuzzy logic.

4.5 Fuzzy logic

Conventional logic is based on the idea that a statement is either true or false. In some cases, this division is valid. A light switch is either on or off, an individual either owns a car or does not. However, generally, this sort of categorisation is alien to the way we think. If you ask whether someone is old or young, or whether some water is hot or cold, the answer will vary for each person who is asked. Typically, an answer will be qualified with statements such as 'quite', 'very', 'slightly', 'almost'. As humans, we understand what is meant. But how can a computer also learn to understand these qualitative terms?

The answer to this question is fuzzy logic. Fuzzy logic was first developed in the 1960s by Lofti Zadeh. He also developed a mathematical framework for it, known as fuzzy set theory. This framework employs a continuously valued logic, utilising all values in the range 0 to 1 and including all the functions available in conventional logic. In fact, conventional set theory can be viewed as a proper subset of fuzzy set theory [10].

4.5.1 Membership

A collection of objects may be considered to be a set. In conventional crisp set theory, an object is either wholly a member of a set or it is not. To use the example of car ownership – someone either owns a car or they do not; they are either in the set of car owners, or they are not. This clearly defined distinction can be illustrated in a Venn diagram (Figure 4.3). In set theory this is described by the principles of membership, where the membership value is a real number in the range of 0 to 1. For crisp sets, if an item is part of a set, it has a membership of 1; if it is not part of the set, it has a membership of 0. There are no intermediate values. So, a person with a car has a membership of 1 of the set of car owners, whereas someone who does not have a car, has a membership of 0 of that set.

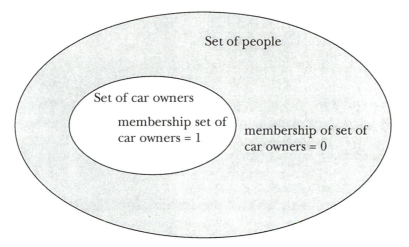

Figure 4.3   Venn diagram showing crisp sets

In fuzzy logic, memberships can take values between these two extremes. This represents the degree to which an item can be considered to be a member of the set. So it is not necessary to decide whether water at 67°C is hot or warm. It can be thought of as being both, to a greater or lesser extent. The same is true for water at 85°C, although that will have a higher membership of hot than the cooler water. This is illustrated in Figure 4.4. A membership set for hot can be generated. It might be decided that 100°C is definitely hot (membership 1 of the set hot) and 50°C is definitely not hot (membership 0 of the set hot). A line may be drawn between these two extremes, representing a linear increase in the membership with temperature. From this, it can be seen that 67°C has a membership of 0·34 and 85°C has a membership of 0·7 of the set hot.

It should be noted that a membership of hot to degree 0·7 is quite different from saying that there is a probability of 0·7 that something is hot. There may be no uncertainty about the temperature, just about the label to attach to that temperature. The fuzzy logic operations described here are identical to a statistical interpretation of membership as 'what is the chance that an independent observer would classify that as being hot?'.

The shape of the membership sets can be triangular, trapezoidal, sigmoidal, Gaussian [2], etc., depending on the application. The shape is an approximation to the distribution of degree of belief that the class truly represents the parameter value presented, i.e. the probability that an observer would classify that temperature as hot. The sets overlap each other to provide a continuous function across the range. Therefore, a value will usually have a membership of more than one set.

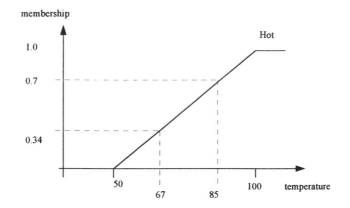

Figure 4.4   Membership of water at 67°C and 85°C of the set hot

4.5.2 Rules

In conventional logic, the effects of the membership of sets are combined using rules based on operators such as AND, OR and NOT. Using these operators, truth tables can be created to give the appropriate output for all the possible input combinations.

Fuzzy logic also uses the logical operators OR, AND and NOT to combine membership sets. However, as the inputs are continuous over the range 0 to 1, it is not possible to generate a truth table. Therefore other methods are used to evaluate the functions. The AND operator may be taken as the minimum of the two input membership values, the OR operator as the maximum of the two input memberships and the NOT operator as one minus the input membership. This is known as max-min inferencing.

Alternatively, the product may be used to evaluate the AND operator, and a method based on the sum minus product for the OR operator. The product version of AND is consistent with the statistical interpretation of membership described in Section 4.5.1. There are a number of other interpretations for the operators that may be found in the literature [6].

4.5.3 Defuzzification

Applying a series of rules will result in a number of different outcomes being fired to varying degrees. The final stage is to translate the fuzzy memberships into a crisp value. This is the value used for the final decision. There are two main methods of defuzzifying the data. These are the centroid (or composite moment) method and the height (or composite maximum) method. Statistically, this is equivalent to choosing different mean weightings of a distribution [7].

The centroid method uses the centroid of the combined output fuzzy sets. This is affected by all the results given by the rules and weights those most strongly activated. The height method uses the average of the output values of the scaled output membership functions, weighted by the heights of scaled membership functions.

4.5.4 Example

Fuzzy logic is a simple, practical method of manipulating uncertain information. Its implementation is best illustrated using an example – in this case a controller for determining the correct speed for driving a car around a bend in the road. In order to calculate the correct speed to drive around a bend, a large amount of accurate data would be required. This would include the angle of the bend, camber of the road, coefficient of friction between the tyres and the road surface and so forth. If a change in speed is required, then a whole further set of calculations arise to determine how much to move the accelerator pedal to give the required acceleration or deceleration. When a driver approaches a corner he or she does not do this detailed calculation. If that level of mathematics was required, then none of us would drive.

Instead we are able to make decisions based on fuzzy logic, using the incomplete and qualitative data available to us. As we approach the corner, we ask, 'am I going too fast?'. If the answer is yes, a little, we might ease off the accelerator a bit. If the answer is yes, much too fast, then we will remove our foot from the accelerator altogether (and probably brake). Our decision is based on a series of rules, taking into account the various data which we have at our disposal.

Imagine that we are approaching the bend. The first stage in our decision making process might be to categorise our current speed as fast, medium or slow. This is illustrated in Figure 4.5. In this case, we have chosen the membership sets to be triangular. If the car speed is 35 mph, this could be considered as being fast to degree 0·25 and medium to degree 0·75.

Having decided, in fuzzy terms, what the speed of the car is, we can perform similar analyses on the other factors. For example, we can classify the bend as sharp, quite sharp, medium, quite gentle and gentle (there is no limit on the number of membership sets), or the road surface as dry, damp and wet.

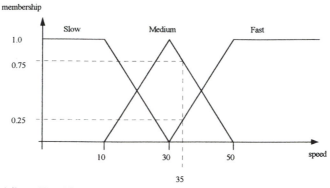

Figure 4.5   Fuzzifying car speed on a bend

Having fuzzified the data, the next stage is to write down a series of rules that describe the required response of the system. Many of these rules are intuitively obvious. For example, for our car example, a rule might be:

IF (speed is fast) AND (bend is sharp) THEN (reduce throttle)

Others might be:

IF (speed is medium) AND (road surface is dry) THEN (keep throttle constant)

IF (bend is sharp) OR (road surface is wet) THEN (reduce throttle)

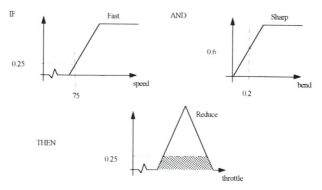

Figure 4.6   Evaluation of a rule using the fuzzy AND operator

To evaluate the first rule, we take the minimum of the membership of 'speed is fast' and of 'bend is sharp'. If 'speed is fast' is true to degree 0·25, and 'bend is sharp' is true to degree 0·6, then the output of the rule will be that 'reduce throttle' is true to degree 0·25. This is illustrated in Figure 4.6. As the other rules are evaluated, then other outcomes will be fired to varying extents.

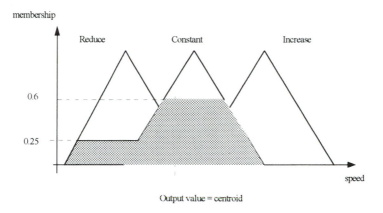

Figure 4.7   Defuzzification

For simple examples it is quite common to use the minimum and maximum operators instead of the more rigorous product and (sum-product) operators. The simplifying assumption is qualitatively correct and the linear membership has already moved the operation away from full rigour.

These outcomes are resolved by defuzzification. There are a number of methods of doing this, e.g. the centroid method illustrated in Figure 4.7. Using this technique, the fuzzy set values are translated into a crisp value. This value can then be output from the fuzzy system.

4.5.5 Advantages of fuzzy logic

There are a number of advantages of fuzzy logic over conventional modelling methods. The IF-THEN-ELSE rule base is easily understood and readily modified by the nonspecialist user. As a result of this, generating the model does not rely on strong knowledge of computer programming or the mathematics that are required by traditional methods. This typically makes a fuzzy logic model faster to develop than a mathematical model. The clear linguistic statements also ensure that the rules used to reach a decision are visible to those using the system, and easily verified.

Fuzzy logic is able to handle highly nonlinear and noisy systems. This can eliminate the problems encountered by many traditional methods by mode-switching phenomena, such as hysteresis, control gain changes, etc.

A good qualitative understanding of the system to be modelled is required in order to develop a good fuzzy system. However, during use, the response of fuzzy systems only degrades gradually as the quality of the knowledge base degrades, making the system very robust.

The response is fast and requires only modest computing power, with a standard desktop PC typically being adequate. Hence, fuzzy logic can run at speeds comparable to those of algebraic equations.

4.5.6 Uses of fuzzy logic

Fuzzy logic has been applied to a range of applications in a wide variety of fields. These include information processing, control, robotics, decision making and support, analysis, diagnosis and prediction. Examples of applications are voice and character recognition; power scheduling; elevator control; industrial plant control; control of focus, exposure and zoom in cameras; car active suspension; stock portfolio management; and information compression.

4.6 Fuzzy logic to resolve constraints in genetic algorithms

Consider the example where a genetic algorithm is being used to obtain a best estimate of the parameters a, b, c and d in the equation:

$$y = a \cos(4x) + b \sin(x) + cx + d$$

and it is known that d lies in the range -100 to 100 and a, b and c lie in the range -300 to 300.

If you have four accurate data points, then these parameters can easily be obtained. However, if only three data points are known, then a range of solutions will fit the data. This is illustrated by doing a number of genetic algorithm optimisations using the data points $(0.1, 11.5241)$, $(0.4, 5.6842)$ and $(0.9, 0.78186)$. The objective was calculated on the basis of least squares, and this range of results is illustrated in Figure 4.8. The solutions are fairly evenly distributed over the set of possible solutions.

However, additional knowledge may be available that further narrows down the set of solutions. This knowledge may not be precise, but a qualitative indication of what is expected. For example, the approximate location of a minimum or maximum, or an approximate relationship between some parameters may be known. This additional information, coupled with the original data, can be sufficient to determine a single solution or small set of potential solutions to the problem.

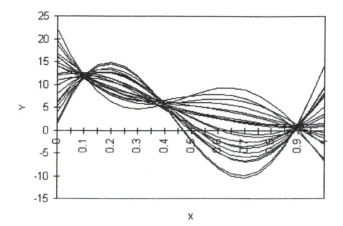

Figure 4.8   Solutions of genetic algorithm search for parameters using only three data points

The quality of the solution in terms of its distance from the known data points can still be calculated using least squares. However, that value can then be fuzzified, by finding its membership of fuzzy sets which describe how well the objective has been met (EXCELLENT, GOOD, OK, BAD and VERY BAD). Triangular, asymmetric sets were used in this case and these are shown in Figure 4.9. This gave a qualitative assessment of how well the solution had achieved its objective of passing through the given data points.

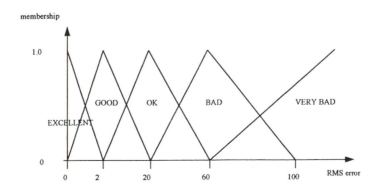

Figure 4.9   Fuzzy sets for quality of match to data points

An example of additional knowledge that may be used to direct the search is that it might be known that the magnitude of parameter b is about twice the magnitude of parameter c. Or, in mathematical terms:

$$|2c| \approx |b|$$

Hence, there are now two aspects to the optimisation. Solutions that pass through or close to the data points are good, but only if the relationship between parameters b and c is as expected. How well the solution met the additional constraint was calculated and turned into a fuzzy description of the degree to which the constraints had been met. These sets are illustrated in Figure 4.10.

A series of rules was then generated that described the way in which the membership values from the two fuzzy sets combined. Examples of these are:

IF (match is excellent) AND (constraint is met) THEN (solution is excellent)
IF (match is excellent) AND (constraint is just missed) THEN (solution is good)
IF (match is very bad) AND (constraint is met) THEN (solution is bad)
IF (match is very bad) AND (constraint is badly missed) THEN (solution is very bad)

The memberships from the fuzzy sets were applied to these rules. The minimum method was used to represent the AND operator. The overlapping fuzzy sets meant that for every solution, multiple rules were fired to some degree. The final solution was obtained by defuzzifying the outputs from the rules by the centroid method.

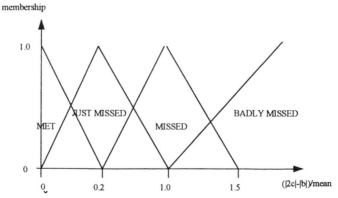

Figure 4.10   *Fuzzy sets for quality of meeting additional constraint (parameter equivalence)*

Using this evaluation, a genetic algorithm was used to generate optimal solutions. The genetic algorithm used was conventional with one-point crossover, mutation and inversion. Real number representation was used for the chromosome strings. Selection was by the roulette wheel method and elitism was applied to the population replacement. A small population of 50 was used in a single population, over 800 generations.

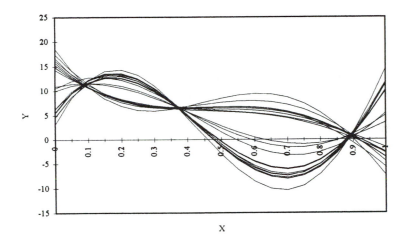

Figure 4.11   Solutions of genetic algorithm search using three data points and additional parameter conditions (parameter equivalence)

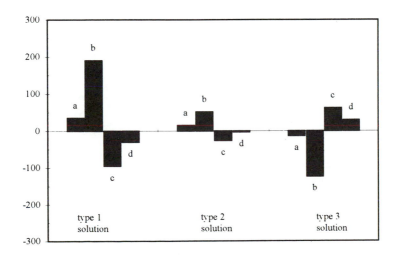

Figure 4.12   Representative solutions of the three clusters

The genetic algorithm was able to generate solutions which passed close to the three data points and with parameters *b* and *c* in the approximate ratios required. The solutions from this fall into three groups. This grouping is visible on the plot of the solutions in *x-y* space (Figure 4.11). It was also illustrated by using a K-mean clustering algorithm [1]. The K-mean method searches for groupings in the data and calculates the mean of each group. The parameters of the equation were used for the clustering. The forms of the solutions in the three groups are shown in Figure 4.12. It can be seen that the three groups are quite distinct from each other. The solutions in each cluster were of similar quality in terms of matching the three data points, and meeting the additional parameter condition.

A similar exercise was performed using the same three data points, this time adding knowledge of the approximate location of a minimum on the curve. The location of the minimum was taken to be between about $x = 0.74$ and $x = 0.76$. As in the previous example, fuzzy sets were used to describe the quality of the solution in terms of its ability to match the data points. The additional requirement was based on the distance of the actual minimum from the required zone (in fact, the sets were taken as the distance from the centre of the required zone, $x = 0.75$). These sets are shown in Figure 4.13.

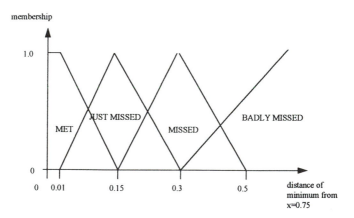

Figure 4.13   Fuzzy sets for quality of meeting additional constraint (location of minimum)

The genetic algorithm was run using the same configuration as before. The solutions generated by the algorithm over several runs are shown in Figure 4.14.

The membership sets and the rules used in this example were not tuned to give optimal performance, but represented a best guess. The fact that the genetic algorithm was able to generate solutions using them illustrates the robustness of the technique.

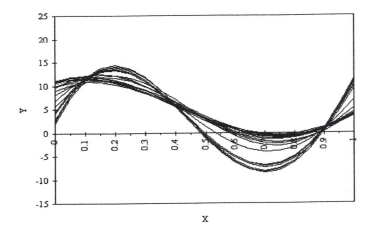

Figure 4.14   *Solutions of genetic algorithm search using three data points and additional parameter conditions (location of minimum)*

4.7 Engineering applications of the technique [9]

In modern gas turbine engines, in order to improve efficiency, the gases passing over the turbine blades are at very high temperatures. This makes it necessary to cool the turbine blade by passing cooler air through a series of passages in the blade. A typical configuration is shown in Figure 4.15. The design of these passages is very important as the airflows affect engine efficiency. The airflows through the passages are modelled using a computer model. However, this model requires calibration against the performance of the actual manufactured blade. Hence, a number of experiments are performed on the blade to determine the actual air flows in the passages. The computer model is then tuned to give results which accurately represent the manufactured blade performance. The parameters used to tune the model are the coefficients of friction within the passages, the discharge coefficients at the outlets and the inlet losses.

Traditionally, this tuning has been done by manually adjusting the model parameters to fit each experiment in turn. However, this is a time consuming process and can result in the solution becoming stuck in a local optimum. Genetic algorithms were considered to be an alternative method of performing the optimisation due to their robustness and ability to avoid local optima. The genetic algorithm used was a conventional algorithm with multiple populations and crossover, mutation and inversion operators [3]. The objective was to match the predicted air mass flows through the passages with those obtained in the experiments.

The results from an initial study were promising. The genetic algorithm repeatedly found solutions with very small deviations from the experimental values. However, over a number of runs, the genetic algorithm located a significant number of different solutions of comparable fitness. This was due to the fact that the problem was underconstrained as a number of criteria used by the designers to determine the quality of a solution had been left out. However, when these were included as hard constraints, the genetic algorithm was unable to find a solution.

Discussions with the designers made it clear that these criteria were guidelines. In an ideal case, all these criteria would be attainable. In practice, noisy data and uncertainty in the model means that engineering judgement is used to determine whether the solution is acceptable.

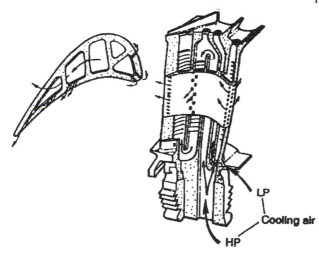

Figure 4.15   Section through a turbine blade showing cooling passages

For example, ideally the model should give a precise match of the experimental data for all configurations and pressures used during the test. However, in practice the engineer applies tolerances on the deviations and the patterns of deviation to determine whether the solution gives a good overall fit or not.

Additional information relates to the relationship between the coefficients applied to adjacent passages. Although these coefficients are unlikely to be exactly the same, they would be expected to be similar. A nominal tolerance on the difference in the coefficients of passages in close proximity may be set to, say, ten per cent. In practice, a slightly higher difference may be acceptable if this gives better matches on the air mass flows. However, it may be acceptable for one pair of rows to lie outside the nominal tolerance, but for all the pairs to lie at the edge of the tolerance bounds is a less likely scenario.

As a result of this, a set of rules was generated to describe the quality of the solution and these were included in the objective function. The genetic algorithm was run a number more times, using these constraints. As before, the solutions had a very good match between the model predictions and the experimental results. However, there was more consistency in the solutions as the constraints served to narrow the search down to regions which gave results that better matched the engineers' judgement of desirable characteristics.

Work is ongoing on this application to extend and refine the rulebase and set definitions.

4.8 Discussion

Fuzzy logic is a well established method for resolving qualitative information in a logical manner. There are a large number of products on the market and in industry that use fuzzy logic, particularly in the area of control.

Fuzzy logic has a number of attributes which could potentially be of benefit in fitness calculations in genetic algorithms.

The linguistic classifications and rules used by fuzzy logic are easily understood. Therefore, the effect of the constraints on the fitness function is visible to the user rather than being obscured in an arbitrary mathematical function. This makes the fitness evaluation routines more easily verified and modified. It also reflects the way that a human would approach the problem.

A fuzzy logic routine does not require long evaluation times. The additional time to perform an analysis is usually insignificant compared with the time to evaluate a typical engineering objective calculation. This is particularly important as the evaluation function is called many times during a genetic algorithm run.

Although the method has only been tested as part of a genetic algorithm search, the fuzzy logic could serve to create a continuous, well behaved search space for other optimisation techniques. Work is continuing to better understand the use of fuzzy logic in this area and its benefits to optimisation in industrial applications.

4.9 Acknowledgments

I would like to thank the following Rolls-Royce personnel for their assistance in the turbine blade work – Graham Purchase and Chris Barnes for technical advice and Chris Bradley for computing support.

I would particularly like to thank Colin Silvester for his advice on the use of fuzzy logic, and Dr P. H. Cowley for his assistance in preparing this Chapter and advice on the techniques.

4.10 References

1   Bishop, C. M.: *'Neural networks for pattern recognition'* (Oxford University Press, 1995)
2   Brown, M. and Harris, C.: *'Neurofuzzy adaptive modelling and control'* (Prentice-Hall International Editions, 1994)
3   Goldberg, D. A.: *'Genetic algorithms in search, optimisation and machine learning'* (Addison-Wesley, 1989)
4   Homaifar, A., Qi, C. X. and Lai, S. H.: 'Constrained optimisation via genetic algorithms', *Simulation*, **62**, (4) pp. 242–254, April 1994
5   Ingber, L.: 'Simulated annealing: practice versus theory', *Math. Comput. Model.* **18**, (11) pp. 29–57, 1993
6   Klir, G. J., and Folger, T. A.: *'Fuzzy sets, uncertainty, and information'* (Prentice-Hall International Editions, 1992)
7   Kosko, B.: *'Neural networks and fuzzy systems'* (Prentice-Hall International Editions, 1992)
8   Kreyszig, E.: *'Advanced engineering mathematics'* (John Wiley & Sons, 4th ed., 1979)
9   Pearce, R., and Cowley, P.H.: 'Use of fuzzy logic to overcome constraint problems in genetic algorithms'. First IEE/IEEE international conference on *Genetic algorithms in engineering systems: innovations and applications*, pp. 13–17, 1995
10 Silvester, C. F., and Ma, K.: 'An introduction to fuzzy logic'. Technical report RR(OH) 1326, Rolls-Royce plc., 1994

Chapter 5
Towards the evolution of scaleable neural architectures

S. Lucas

5.1 Introduction

There has been a great deal of interesting work published on evolving neural networks in the last few years, some of which is mentioned below. Nearly all previous work, however, has concentrated on evolving a particular neural network to solve a particular problem. When a suitable solution is evolved, then all we have is a suitable solution for a particular problem. The work reported here offers a significant departure from that theme, and presents a simple system which allows the evolution of scaleable neural architectures.

This is important for two reasons:

- Evolutionary search is computationally expensive. When evolving solutions to complex problems, it might be better to evolve solutions to small examples of the problem, then for the real application, scale up some of the best evolved solutions to the real problem size.
- Having evolved good solutions to a problem, it would be good to apply them to similar problems of a different size. By allowing this, we allow maximum possible reuse of neural network modules.

There are two obvious ways in which this sort of scaleable reuse may be achieved. The first way is through iterated modularity. That is, a particular module that performs a specific task is used many times. For example, in a neural network computer vision system, it would be appropriate to employ identical edge detection modules at regular intervals in the input space. The second way we may achieve scaleability is through learning: instead of encoding in the chromosome a solution to a specific problem, we can instead encode a general architecture which is able to learn solutions to problems for itself. There is a parallel here with the different kinds of behaviours observed in living organisms. Generally speaking, simple behaviours can be genetically preprogrammed into the organism, whereas more complex behaviours have to be learned during the life of the organism — for this, the organism can only be genetically preprogrammed with the ability to learn these behaviours.

A neural architecture is a class or type of neural network, such as a perceptron or multilayer perceptron. Many neural architectures have more than one mode of behaviour; note that to completely specify an architecture, we must fully specify each mode of behaviour. For example, all supervised feedforward neural architectures have two modes of behaviour: a training mode and a recognition mode. During training, an input vector is presented to the network, and the network uses its current setup to compute an output vector. This is the feedforward part. The next step is to apply the supervisory signal (target vector), calculate the error between the output and target vectors and compute updates to the network parameters in order to minimise this error. This step is the feedback part. To completely specify a supervised feedforward architecture, we should specify both its feedforward (recognition) and feedback (training) behaviours. The main result of this Chapter is that given the feedforward part of a neural network, we can evolve the feedback part, and hence its learning behaviour.

The key to the successful evolution of scaleable neural architectures is the encoding scheme i.e. how the neural network is represented in the chromosome. The potential benefits of an evolutionary approach in conjunction with a good encoding are:

- Current neural architectures are limited in application by the ability and imagination of their designers; an evolutionary approach is not necessarily bounded by such limitations.
- Evolutionary techniques can be used to select the best architecture for a given problem, as well as the details of that architecture.

5.2 Encoding neural networks in chromosomes

There are two distinct methods of encoding neural networks in chromosomes: direct and indirect. These are depicted in Figures 5.1 and 5.2, respectively.

In the direct case, we can identify a direct correspondence between each part of the chromosome and each part of the network. There are still many possible ways of directly specifying a network. The chromosome can directly specify the network structure and the network weights, or it can be used to specify just part of the network. For example, given the exact structure of a multilayer perceptron, i.e. the number of layers and the number of units in each layer, we can encode all the weights in a chromosome and use a GA to evolve a set of weights which solves a particular problem. Another example of a direct encoding would be a totally general description of a network which simply listed the function of each node (with possibly a variable number of nodes) and the connection weights between each node. The only constraint that makes an encoding direct is that there exists a one-to-one correspondence between each element of the chromosome and each variable element of the network.

The indirect encoding method is where no such correspondence exists. All indirect encodings that have been presented in the

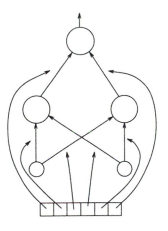

Figure 5.1 A direct encoding scheme

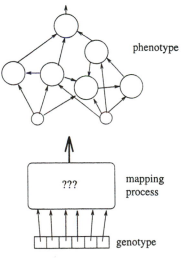

Figure 5.2 An indirect encoding scheme

literature have the property that small chromosomes can generate large networks. The argument is then often made that, since the size of the search space has been reduced, so a GA is more likely to find a good solution. In fact, the size of a search space alone is no indicator of how easy that space is to search – the shape of the space is also important.

Evolutionary methods have been used most extensively in conjunction with a direct encoding, and there are hundreds of papers in the literature that have adopted this approach. Some have used a fixed architecture, and evolved only the weights of that architecture [1, 2], others have evolved both the structure and the weights [3, 4]. Interesting neural solutions have been evolved to problems in time series prediction [5, 6], playing tic-tac-toe [2] and in evolving dynamical networks which learn to output simple sequences [1]. This latter approach is of particular relevance to the work reported here, since a learning behaviour was evolved. However, due to the use of a direct encoding, the learning behaviour is not transferable to other problems of a different size. Furthermore, although the direct encoding method has the virtue of simplicity it scales poorly, and hence is only suited to evolving relatively small networks.

For this reason there has been some interesting work done on indirect encoding, where the chromosome is used to drive some kind of network construction algorithm. Such methods have the property that small chromosomes can grow massive neural networks.

Some of the earliest work in this area was reported by Kitano in 1990 [7], and has since been followed up with superior network generation languages and grammars designed by Gruau [8, 9], Boers and Kuiper [10], Muhlenbein and Zhang [11, 12] and Sharman *et al.* [13]. All of these, however, either use the GA (operating on strings or graphs in the neural description language or chromosome) to evolve a hard-wired neural network, or use the GA to evolve a good topology (or evolve a good topology and weight set) which is then trained by error back propagation or simulated annealing.

Kitano [14] has also developed a unified framework in which the network structure and weights are allowed to evolve, and followed this up with a more biologically detailed simulation [15]. Even in [14], however, despite claims that all details of the network are allowed to evolve, this is not quite true, since the learning algorithm is fixed in advance to be error back propagation – although the network topology and initial weights are evolved.

In contrast to this, the author [16, 17, 18] has shown how it might be possible to evolve the learning algorithm within the same unified framework in which the other network details are evolved. The work reported here goes one step further than the previous work by actually demonstrating the evolution of learning behaviour for a perceptron-type network, and then showing that the learning behaviour evolved for learning one problem can be subsequently reapplied to a completely different problem. There are two distinct features of the evolutionary framework reported here that make this possible and practical: the active weight model and a set-based chromosome structure. These are discussed below.

Related work of interest is that of Montana [19] with his strongly typed genetic programming (STGP), which he uses to evolve programs that explicitly operate on vector, matrix and list data structures, as well as simpler data types. Within this framework he evolves (among other things) the algorithm for updating the input tracking estimate in a Kalman filter.

Finally, on the subject of encoding, we present some ideals which artificial chromosomes used for neural architecture evolution should aspire to:

- The chromosome should be modular. This will allow evolution to proceed by putting together existing building blocks in new ways as well as developing new building blocks.
- The chromosome should be human readable – after spending a good deal of time evolving a novel solution to an interesting problem, it would be a shame if we could only appreciate it at the connection-matrix level.
- The chromosome should be parameterised. Hence, having evolved a new type of architecture for a class of problem, we should be able to parametrically alter it to tackle related problems of a different size.

5.3 Evolutionary algorithms

Some grand claims have been made for the ability of genetic algorithms to perform efficient search [20]:

> 'Indeed, the number of strings in a given region increases at a rate proportional to the statistical estimate of that region's fitness. A statistician would need to evaluate dozens of samples from thousands or millions of regions to estimate the average fitness of each region. The GA manages to achieve the same result with far fewer strings and virtually no computation.'

Such claims are terribly optimistic, and the 'no free lunch' theorem of Wolpert and Macready [21] offers grounds for treating such claims with a good deal of caution. This states that when averaged over the space of all possible cost functions, there is no reason to prefer one search algorithm to another. In other words, there is no way that the GA could possibly offer this super efficient search in all search spaces.

The important question, of course, is which algorithms perform best on the kind of search spaces that we encounter when solving problems of interest to us.

The key point of evolutionary search algorithms is that they base the search only on evaluation of the cost function, and pay no attention to the intrinsic properties of the models being evolved. There are two main branches of evolutionary algorithms: random hill climbers, and population based search methods (most of which use crossover to combine individuals) – the latter most commonly known as genetic algorithms [22]. The random hill climber is a special case of simulated annealing [23, 24] at zero temperature. Despite the claims of Holland [20], most studies have either found little preference for one or the other type of algorithm [25], or found that the random hill climber performs best most of the time [26, 27]. The big problem with random hill climbing strategies is that they can get stuck in deep local minima. In [27] a dynamically changing representation was used to overcome this problem – this is essentially similar to the macro-mutation model of Jones [28]. Both methods offer the possibility of (effectively) taking large steps to escape local minima. Culberson [29] devised several cost functions, and showed some that favoured random hill climbing mutation only, and others that favoured crossover.

Also, there have been some studies regarding what makes a problem hard for a GA to solve. Jones and Forrest [30] describe the fitness distance correlation measure as a reasonably reliable indicator of how hard a problem will be for a GA to solve. Altenberg [31] reports that one of the keys to a GA being successfully deployed is the fitness correlation between parents and their offspring.

Regarding the training of neural networks, some authors have reported that in many cases random hill climbing (or GA with mutation only) simply fails to find a solution on many problems, even as simple as XOR. For example, Robbins *et al.* [32] found that crossover was necessary to evolve weight sets for an MLP to solve the XOR problem. More recent work has offered counter evidence to this type of claim [33, 34].

In the experience of the author, error back propagation with multiple random starts nearly always outperforms evolutionary search for sets of optimal weights for particular problems (such as XOR, spirals or pattern recognition data), in terms of CPU time required to find a solution of a given quality. Hence, it is of interest to evolve not just a solution to a specific problem, but to evolve a network that can learn to solve many kinds of problem.

In general, it seems that the choice of representation and the choice of operators on that representation play a more important role than whether a population based GA or a random hill climber is used as the search algorithm. A classic example of this is the travelling salesman problem. Whether we use a random hill climber or a population based GA to solve it is largely immaterial – the most critical effect on the efficiency of the search is that an inversion operator be used to macro-mutate the routes, as used by Lin [35] for example.

5.4 Active weights and the simulation of neural networks

This Section introduces an alternative way of viewing a neural network, called the active weight model. This is the key to evolving neural networks that have intrinsic learning behaviours. The traditional passive weight model is depicted in Figure 3. Here the active computing units are connected by passive connection weights which have no way of modifying themselves. This leads to the standard view of neural networks, such as multilayer perceptrons, where the learning algorithm is separated from the actual network.

The active weight equivalent network of the one passive weight net in Figure 5.3 is shown in Figure 5.4. Under the active weight model, all connections between neurons are unity – traditional connection weights can be modelled by feeding the weight value together with the cell output into a product neuron, and connecting the output of the product neuron to the input of the summation cell.

The big advantage of the active weight model is that we can now simulate any neural network with one simple algorithm:

repeat (for life of network)
$$\forall i \in N$$
$$o_i := f_i(I_i)$$

- N is a set of neurons.
- ith neuron computes function f_i which maps from the set of input values I_i to an output value o_i.

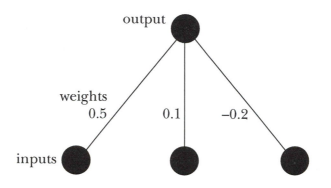

Figure 5.3   *The standard way of viewing a neural network: the neurons (represented by shaded circles) are the active computation elements and the weights connecting them are passive, requiring a separate algorithm to update them*

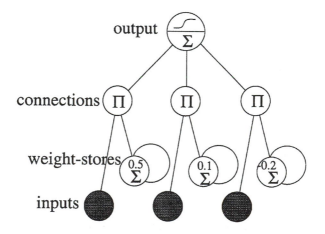

Figure 5.4   *The active weight view of a neural network. All connection weights are now unity and remain so throughout the life of the neural network. By utilising product neurons, however, the output values of some ordinary neurons can now behave just like connection weights – these weight-store neurons are shown here with values to make this network behave just like the one shown in Figure 5.1. The advantage is that, under this strategy, learning behaviour becomes an intrinsic property of the neural network and does not require a separate algorithm*

An active weight learning perceptron is depicted in Figure 5.5. This has the least mean squares learning rule built into it, i.e. when simulated by the above algorithm this network performs gradient descent to minimise the LMS error between the output cells and the target cells.

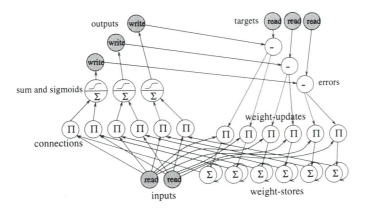

Figure 5.5   *The learning perceptron. The shaded neurons read from the input stream or write to the output stream. Connections to the threshold unit are not shown in order to simplify the diagram. The Σ and Π units compute the sums and products of all their inputs respectively. This perceptron trains itself using the delta rule, provided that the cells are updated set by set working clockwise from the inputs to the weight stores*

The import of this is significant; the active weight model allows us to in principle evolve learning algorithms, just by specifying network topologies. We now have to concentrate our efforts on efficient ways of specifying complex topologies.

5.5 A set based chromosome structure

This Section describes one possible approach towards efficiently specifying (and hence evolving) all aspects of a neural network. It is just an initial step down this road – future specification languages will need to be more explicitly modular and offer different modes of behaviour for the network.

The set-based chromosome is split into four distinct parts. For a given problem some of the genes within a chromosome must be fixed. In effect, this allows the search for a fit individual to be concentrated within a particular species of network. By fixing certain genes we can ensure that all chromosomes generated will give rise to networks which at least have a structure which is viable for solving the problem. It is a waste of CPU time to bother looking at networks that have the wrong number of outputs, for example.

The first part consists of a list of size declarations. The value of each size expression is allocated to successive integers.

A list of set declarations comprises the second part. Each set is declared as being of a given type, and of a given size. The size is specified by an integer which refers to a size declaration from the first part of the chromosome. The types allowed at the moment permit summation, product, difference, read, sigmoid, constant and sumstore neurons (which sum all their inputs together with their previous value). Also, there is a mergeset, with the purpose of simply amalgamating two other sets.

The third part is a list of connections between sets, whose nature is described below.

The fourth part of the chromosome is a list of input/output connections for any phenotypes grown from this chromosome. Without this part the created networks would be isolated from any interaction with the system in which they are evolved.

5.5.1 Set interconnections

At present, set connections may be one of two types, called Div and Mod, after the operators used to calculate which node in set **a** to connect to which node in set **b**. Figure 5.6 illustrates this. Set **a** is a sumset of size 3, set **b** is a sumset of size 2, set **c** is a prodset of size 6 and set **d** is a sumset of size 2. The interconnections are as follows: (**c a** Mod) (**c b** Div) (**d c** Div).

By appropriate use of Div and Mod connection specifiers in conjunction with sets of arithmetic operators, any of the standard vector/matrix operations are possible, plus many other nonstandard ones. For example, the set **c** represents $c = ab^{-1}$ in standard vector notation.

5.5.2 Example chromosome

We now look at an example chromosome template for evolving a single layer perceptron with learning behaviour (see Table 5.1). Most of the chromosome for this network has to be fixed to ensure that we search a viable space of networks.

Table 5.1 The four segments of the chromosome. The first part declares the possible sizes of each set, the second part declares the composition of each set, the third part declares the groups of connections between neurons of each set, and the fourth and final part declares the external connections

Size Def	Parameter name
0	inputs
1	1
2	outputs
3	1+inputs
4	(1+inputs) *outputs

Set	Function	Size Var	Comment
0	readset	0	the set of inputs
1	constant	1	a set with a single constant unit
2	mergeset	3	the augmented input set
3	readset	2	target vector
4	prodset	4	connections
5	accset	2	synaptic activity at output
6	sigset	2	sigmoid of synaptic activity
7	diffset	2	difference between output and target
8	any type	{0,1,2,3,4}	randomly chosen set type and size
9	sumstore	4	the weight stores

Connection	To	From	Type
0	2	0	Mod
1	2	1	Mod
2	4	2	Mod
3	4	9	Mod
4	5	4	Div
5	6	5	Mod
6	7	3	Mod
7	7	6	Mod
8	8	any set	{Mod, Div}
9	8	any set	{Mod, Div}
10	9	any set	{Mod, Div}

External Name	Set number
inputs	0
targets	3
outputs	6

We have a set of *I* inputs which are augmented with a constant unit that always outputs 1·0, to allow each output to have a different bias depending on the weight connecting it with the constant unit. Therefore, if we have *O* outputs, we need a set of *W* weights where *W* = *O* × (1 + *I*). This is shown in the first part of the chromosome.

Hence, any viable solution network must include sets of these sizes. For the next part we must declare some necessary sets, and allow other sets (in fact just one in this case) to be chosen randomly. The following set declarations are commented on to aid understanding.

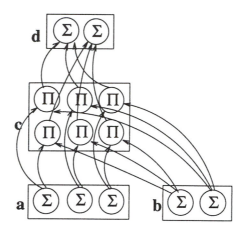

Figure 5.6 Illustration of how the set connection operators work

The only set definition that is subject to evolutionary change is that of set 8. This was chosen because a solution was known to exist using this number of sets. For future work, more open-ended experiments are intended. The third part of the chromosome shows the groups of connections between sets. At the moment, each set has an arity of zero, one or two. Readsets have an arity of zero, indicating that they must not take inputs from any other sets in the network – this would destroy their purpose, which is to take inputs from sources external to the network. Accsets (accumulator set) and Sumstores each have an arity of one. Accsets are used to define a set of neurons whose purpose is each to accumulate a sum of their inputs. Sumstores are identical except that they also include in the summation their previous output. The set of connections is also mostly fixed, with the only variability allowed in connections 8, 9 and 10.

Finally, the connection between the neural network and the system that calls it is set up. Most supervised networks can be seen in black box terms as taking information from the outside world via their inputs and their targets and returning information via their outputs. The final part of the chromosome simply specifies which sets perform these input/output functions.

Note that, by fixing many of the details of the chromosome in this way, we are making the actual search space tiny relative to that explored by most evolutionary algorithms. The work presented here is just a demonstration of a concept – larger search spaces will be tackled in future work.

5.5.3 Results

Initial results demonstrate that learning behaviour can be evolved within this framework. The aim was to evolve networks that could learn simple linearly separable functions such as AND and NOR etc. The chromosome template discussed above was used to generate a random chromosome. This was then adapted using a random hill-climbing procedure, or just used in a pure random search. Future work will also involve experimentation with conventional population GAs, although it should be noted that these do not necessarily outperform random hill-climbing methods.

The fitness function was simply the mean square error averaged over one complete pass through all the data (i.e. one epoch), measured just once after 12 epochs through the AND dataset. Hence, the fittest possible individual would have a score of zero. The fitness of the delta rule was found to be 0·009, and the best individual found by the random hill climber had a fitness of 0·004. Figure 5.7 shows the behaviour of some individuals chosen from one run of the algorithm, including the best one that was found in this particular case after 9718 steps. A network was considered to have learnt the problem if its average squared error fell below 0·01 for the final epoch. All the networks that were able to learn the AND data, were also able to learn the NOR data to a similar degree. This proved that we had evolved networks that could learn datasets of a given size and nature, and not just evolved a network for a specific dataset.

One interesting and slightly worrying aspect of the results is that, so far, pure random search has outperformed random hill climbing – although it is too early to draw any concrete conclusions on this point. Given ten runs of the experiment, allowing a maximum of 10 000 fitness evaluations, the random hill climber only found a solution two times out of ten (but did so with a mean of 289 steps). Pure random search found a solution every time within 10 000 evaluations, with a mean of 2152 fitness evaluations. This is probably due to the highly discontinuous nature of the search space that follows from using set-based chromosomes, in that changing the function of a set or the source set of one of its inputs typically has a dramatic effect on network behaviour.

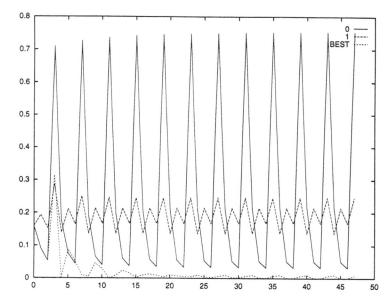

Figure 5.7   *Plots of mean square error with respect to cycle number for some randomly chosen networks (0 and 1), together with the best one that was evolved (best). Each cycle represents a complete update of the network, with a new input/target pair being presented for each cycle. There were four patterns in the data set, representing the AND function – this accounts for the periodicity of length four. Each network was run for 12 epochs – hence 48 cycles in all. Note that for the best network, the error quickly becomes very small for all the patterns – indicating that it has learned the data well*

5.5.4 Scaleability

The claim is that since the chromosomes are parameterised, all that, we need to do to apply the evolved chromosome to a problem of a different size is to change the values of the relevant parameters – in this case, the number of inputs and the number of outputs. This was tested as follows.

Some of the best solutions to the above problem i.e. networks with behaviours which led them to learn simple Boolean functions were then applied to a different type of problem (an OCR problem) of a different size: recognising handwritten digits, where the inputs were a 16-dimensional radial mean vector [36, 37], and the output was a 1-out-of-*N* coding of the character class (hence, ten-dimensional output vector). The data set contained 1669 characters distributed between the ten classes. Here we just quote results on this dataset. A 1-nearest-neighbour weighted Euclidean classifier recognised this set with 77 % accuracy. Normally, of course, we would quote results on some standard database, and use a disjoint training and testing set, but here the only purpose was to demonstrate that the network could learn, to some extent, a completely different problem to that for which it had been evolved.

Most of the networks that were able to learn the Boolean function failed dismally to learn the OCR data—some getting around 12 % correct, others getting only about five characters out of 1669 correct —thus having learned to perform particularly badly at the problem (since we would expect to get around 167 (i.e. ten per cent) correct just by outputting the same thing all the time, or randomly guessing). This poor performance was either because they had a set with an incorrect size variable, or because the connection types were wrong. Both of these types of error can produce network configurations which happen to be correct for problems of a particular size, but that are not correct in the general case. This sort of problem can be averted by including problems of different sizes in the set of problems on which the networks are originally evolved.

Some of the networks, however, were able to learn the OCR data to a reasonable degree – correctly classifying 60 % of the characters. This shows that a general learning behaviour had been evolved – one applicable to a variety of problems.

5.6 Conclusions

Given the structure of a single-layer perceptron as a starting point, a learning behaviour has been evolved. The chromosome that describes the network is parameterised. This allows the chromosome to specify neural networks for other problems of a related nature, but requiring a different size of neural network. Future work will concentrate on evolving learning behaviours for more complicated, more useful networks, and also in evolving new classes of neural networks, together with their learning behaviours. Although this is not the first time that learning behaviours for neural networks have been evolved (see, for example [38, 39]), the manner in which it has been done is more elegant, unified and, most importantly, more open ended than previous approaches, where the nature of the learning rule has been fixed in advance and only its parameters optimised.

The use of a set-based chromosome goes some way towards making the evolved solution well structured and amenable to human understanding. This is hardly an issue with the simple types of network evolved in this Chapter, but will become more important as more complex and interesting architectures are evolved. To this end the author is currently working on a modular specification language (i.e. chromosome language) which will be more readable than the set-based chromosome, and will also allow network modules to have different behaviours – this allows us to conveniently run the network in nonlearning mode when appropriate. Initial calculations indicate that by using this scheme it will be possible to evolve learning behaviours for multilayer networks such as multilayer perceptrons and radial basis function networks – plus many other kinds of neural architecture not yet dreamed of.

5.7 Acknowledgment

This work was supported by UK EPSRC grant GR/J86209.

5.8 References

1 Yamauchi, B., and Beer, R.: 'Sequential behaviour and learning in evolved dynamical neural networks', *Adapt. Behav.*, **2**, pp.219–246, 1994

2 Fogel, D.: 'Using evolutionary programming to create networks that are capable of playing tic-tac-toe'. Proceedings of IEEE international conference on *Neural networks*, San Francisco, 1993, pp. 875–880

3 Dasgupta, D., and McGregor, D.: 'Designing application specific neural networks using the structured genetic algorithm'. Proceedings of COGANN-92 – IEEE international workshop on *Combinations of genetic algorithms and neural networks*, Baltimore, 1992, pp. 87–96

4 Marti, L.: 'Genetically generated neural networks II: searching for an optimal representation'. Proceedings of the international joint conference on *Neural networks (Baltimore '92)*, San Diego, CA, 1992, pp. I, 221–226

5 McDonell J., and Waagen, D.: 'Neural network structure design by evolutionary programming'. *Proceedings of the third annual conference on*

evolutionary programming, Fogel, D., and Atmar, W. (eds.), (Evolutionary Programming Society, 1993) pp. 79–89

6   McDonell, J., Page, W., and Waagen, D.: 'Neural network construction using evolutionary search', in *Proceedings of the third annual conference on evolutionary programming*, Sebald, A., and Fogel, L. (eds.), (World Scientific, 1994) pp. 9–16

7   Kitano, H.: 'Designing neural networks using genetic algorithm with graph generation system,' *Complex Syst.* **4**, pp. 461–476, 1990

8   Gruau, F.: 'Cellular encoding of genetic neural networks,' Laboratoire de l'Informatique du Parallelisme technical report 92-21, Ecole Normale Superieure de Lyon, 1992

9   Gruau, F.: 'Automatic definition of modular neural networks,' *Adapt. Behav.*, **3**, pp. 151–183, 1994

10 Boers, E., and Kuiper, H.: 'Biological metaphors and the design of modular artificial neural networks.' Masters thesis, Department of Computer Science and Experimental and Theoretical Psychology, Leiden University, the Netherlands, 1993

11 Muhlenbein, H., and Zhang, B.: 'Synthesis of sigma-pi neural networks by the breeder genetic programming,' in Proceedings of IEEE international conference on *Evolutionary computation*, Orlando 1994, pp. 318–323

12 Zhang, B., and Muehlenbein, H.: 'Balancing accuracy and parsimony in genetic programming'. *Evolutionary Computation*, **3**, pp. 17–38, 1995

13 Sharman, K., Esparcia-Alcazar, A., and Li, Y.: 'Evolving signal processing algorithms by genetic programming'. Proceedings of IEE 1st international conference on *Genetic algorithms in engineering systems: innovations and applications*, London: 1995, pp. 473–480

14 Kitano, H.: 'Neurogenetic learning: an integrated model of designing and training neural networks using genetic algorithms', *Physica D,* **75**, pp. 225–238, 1994

15 Kitano, H.: 'A simple model of neurogenesis and cell differentiation', *Artificial Life*, **2**, pp. 79–99, 1995

16 Lucas, S.: 'Growing adaptive neural networks with graph grammars'. Proceedings of European symposium on *Artificial neural networks (ESANN '95)*, Brussels, 1995, pp. 235–240

17 Lucas, S.: 'Towards the open-ended evolution of neural networks'. Proceedings of IEE 1st international conference on *Genetic algorithms in engineering systems: innovations and applications*, London: 1995, pp. 388–393

18 Lucas, S.: 'Evolving neural network learning behaviours with set-

based chromosomes'. Proceedings of European symposium on *Artificial neural networks (ESANN '96)*, Brussels, 1996, pp. 291–296

19 Montana, D.: 'Strongly typed genetic programming'. *Evolutionary Computation*, **3**, pp. 199–230, 1995

20 Holland, J.: 'Genetic algorithms', *Sci. Am.*, July, 1992

21 Wolpert, D., and Macready, W.: 'No free lunch theorems for search'. Santa Fe Institute technical report SFI-TR-95-02-010, 1995

22 Goldberg, D.: *Genetic algorithms: in search, optimisation and machine learning* (Addison Wesley, 1989)

23 Metropolis, N., Rosenbluth, A., Rosenbluth, M., Teller, A., and Teller, E.: 'Equations of state calculations by fast computing machines', *J. Chem. Phys.*, pp. 1087–1092, 1953

24 Kirkpatrick, S., Gelatt, S., and Vechi, M.: 'Optimisation by simulated annealing', *Science*, **220**, pp. 671–680, 1983

25 Ackley, D.: 'An empirical study of bit vector function optimisation', in *Genetic algorithms and simulated annealing*, Davis, L. (ed.), (Morgan Kaufman, 1987), pp. 170–204

26 Mitchell, M., Holland, J., and Forrest, S.: 'When will a genetic algorithm outperform hill climbing?' in *Advances in neural information processing systems 6*, Cowan, J., Tesauro, G., and Alspector, J. (eds.), (Morgan Kaufman, 1994), pp. 51–58

27 Kingdon, J., and Dekker, L.: 'The shape of space'. Proceedings of IEE 1st international conference on *Genetic algorithms in engineering systems: innovations and applications*, London, 1995, pp. 543–548

28 Jones, T.: 'Crossover, macromutation and population-based search', in *Proceedings of the sixth international conference on genetic algorithms*, Eshelman, L. (ed.), (Morgan Kaufman, 1995), pp. 73–80

29 Culberson, J.: 'Mutation-crossover isomorphisms and the construction of discriminating functions', *Evolutionary Computation*, **2**, 1994, pp. 279–311

30 Jones, T., and Forrest, S.: 'Fitness distance correlation as a measure of problem difficulty for genetic algorithms', in *Proceedings of the sixth international conference on genetic algorithms*, Eshelman, L., (ed.), (Morgan Kaufman, 1995), pp. 184–192

31 Altenberg, L.: 'The evolution of evolvability in genetic programming', in *Advances in genetic programming*, Kinnear, K. (ed.), (MIT Press, 1994)

32 Robbins, G., Plumbley, M., Hughes, J., Fallside, F., and Prager, R.: 'Generation and adaptation of neural networks by evolutionary techniques (gannet)', *Neural Computing and Applications*, **1**, pp. 23–31, 1993

33 Fogel, D., and Stayton, L.: 'On the effectiveness of crossover in simulated evolutionary optimisations', *BioSystems*, **32**, pp. 171–182, 1994

34 Porto, V., Fogel, D., and Fogel, L.: 'Alternative neural network training methods', *IEEE Expert*, June, pp. 16–22, 1995

35 Lin, S.: 'Computer solutions of the travelling salesmen problem', *Bell Syst. Tech. J.*, **44**, p. 2245, 1965

36 Yamamoto, K., and Mori, S.: 'Recognition of hand-printed characters by an outermost point method', *Pattern Recognit.*, **12**, pp. 229–236, 1980

37 Lucas, S., Vidal, E., Amiri, A., Hanlon, S., and Amengual, A.: 'A comparison of syntactic and statistical techniques for off-line ocr', in *Lecture notes in artificial intelligence (862): grammatical inference and applications*, (Springer-Verlag, 1994), pp. 168–179

38 Bengio, S., Bengio, Y., and Cloutier, J.: 'Use of genetic programming for the search of a new learning rule for neural networks'. Proceedings of IEEE international conference on *Evolutionary computation*, Orlando, 1994, pp. 324–327

39 Chalmers, D.: 'The evolutions of learning: an experiment in genetic connectionism', in *Proceedings of the 1990 connectionist models summer school*, Touretzky, D., Elman, J. Sejnowski, T., and Hinton, G. (eds.), (Morgan Kaufman, 1990)

Chapter 6
Chaotic systems identification
R. Caponetto, L. Fortuna, M. Lavorgna, G. Manganaro

Increasing interest has recently been devoted to the synchronisation of nonlinear circuits (SNC) with particular attention to the case of chaotic circuits [1–4]. Some techniques have been developed for forcing two (or more) identical non-linear dynamic circuits, starting from different initial conditions, to synchronise, namely to follow identical trajectories (asymptotically, at least). This is particularly interesting if the circuits behave chaotically because of their sensitive dependence on initial conditions. However, if some conditions on the so-called conditional Lyapunov exponents are satisfied [1–2] then these circuits can be successfully synchronised. Synchronisation principles have been applied to realising analogue masking systems for secure communication [2]; however SNC is a very promising subject with potential applications to nonlinear systems, signal processing and control.

In this Chapter a new application for the synchronisation techniques is introduced. They have been used to estimate the unknown parameters of a nonlinear chaotic dynamic circuit whose mathematical model is known. In particular, the Pecora-Carroll system decomposition and the cascaded synchronisation approach have been considered [1–3]. A circuit with unknown parameters is used as master in a synchronisation set up and another circuit, with the same mathematical model, is used as slave to estimate the parameters. Identification has been formulated as an optimisation problem: the parameters of the slave have to be chosen in order to minimise a performance index defined so that it has a minimum if the two circuits are synchronised, that is, if they have the same parameter set. Optimisation of the performance index has been efficiently accomplished by means of a genetic algorithm (GA) [5–6]. This method has been used to estimate the parameters of Chua's oscillator [7] which is known for its diversity of dynamic behaviour. In particular its dimensionless state equations have five dimensionless parameters corresponding to seven circuit parameters [8]. The proposed approach has been used to estimate all of these dimensionless parameters.

In the following Sections, some background concepts are first summarised, then the new synchronisation based identification approach is introduced. Finally, some examples, corresponding to different attractors from Chua's oscillator, are presented.

6.1 Background

In this Section Chua's oscillator, the Pecora-Carroll synchronisation technique and the basic philosophy of genetic algorithms are briefly recalled. The main concepts and terminology are discussed.

6.1.1 Chua's oscillator

Chua's oscillator has been extensively studied because of its extremely wide variety of dynamic behaviour together with its relatively simple mathematical model [7]. The oscillator is composed of two capacitors, one inductor, two resistors and a piecewise linear resistor, as shown in Figure 6.1, and is easily realisable.

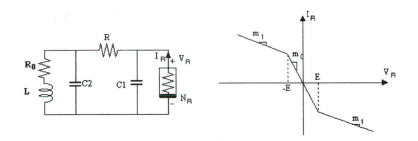

Figure 6.1 Chua's oscillator

It is considered to be a canonical circuit [8] and a benchmark for studying nonlinear dynamics. Its dimensionless state equations are:

$$\dot{x} = \alpha(y - l(x))$$
$$\dot{y} = x - y + z \qquad (6.1)$$
$$\dot{z} = -\beta y - \gamma z$$

with

$$l(x) = m_1 = 0.5(m_0 - m_1)(|x + 1| - |x - 1|) \qquad (6.2)$$

x, y and z being the state variables and α, β, γ, m_0, m_1 the five dimensionless system parameters.

The well known double scroll attractor is observed in Chua's oscillator if:

$$\alpha = 9,\ \beta = 14.286,\ \gamma = 0,\ m_0 = -1/7,\ m_1 = 2/7 \qquad (6.3)$$

Its phase portrait in the x–y plane is shown in Figure 6.2. Two other different attractors from Chua's oscillator are shown in Figures 6.3 and 6.4 respectively; these have been obtained by using the two following sets of parameters [7]:

$$\alpha = -4.08585,\ \beta = -2,\ \gamma = 0,\ m_0 = -0.142857,\ m_1 = 0.285714 \quad (6.4)$$

$$\alpha = 6.579,\ \beta = 10.989,\ \gamma = -0.0447,\ m_0 = -0.18197,\ m_1 = 0.3477 \qquad (6.5)$$

In the following these two attractors will be referred to as attractor no.1 and attractor no. 2, respectively.

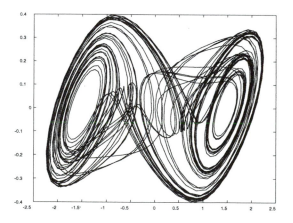

Figure 6.2 Phase portrait of the double scroll attractor in the x-y plane

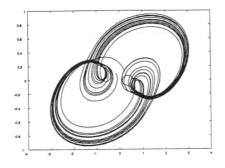

Figure 6.3 Phase portrait of attractor no. 1 from Chua's oscillator in the x–y plane

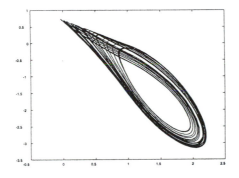

Figure 6.4 Phase portrait of attractor no. 2 from Chua's oscillator in the x–z plane

6.1.2 Synchronisation of nonlinear systems

Synchronisation of nonlinear systems is defined as follows [2].

Definition 6.1: Let us consider two (or more) nonlinear systems ($N{\geq}2$):

$$\dot{x}_i = f_i(x_i) \tag{6.6}$$

where $x_i \in \mathcal{R}^n$, $f_i : \mathcal{R}^n \to \mathcal{R}^n$ and $1{\leq}N$, if:

$$\lim_{t \to \infty} |x_i(t) - x_j(t)| = 0 \tag{6.7}$$

with $i{\neq}j$ N systems are synchronised.

In this Chapter just two systems will always be considered, $N = 2$; these are usually called the master and the slave systems. Many different techniques have been developed to obtain the synchronisation of these two systems [2]; one of these is the Pecora-Carroll approach [1–3], with which a nonlinear dynamic system is considered:

$$\dot{u} = f(u) \tag{6.8}$$

This is partitioned into two subsystems, $u = u(v, w)$:

$$\begin{aligned} \dot{v} &= g(v, w) \\ \dot{w} &= h(v, w) \end{aligned} \tag{6.9}$$

where $v = (u_1, \ldots u_m)$, $g = (f_1(u), \ldots f_m(u))$, $w = (u_{m+1}, \ldots, u_n)$, $h = (f_{m+1}(u), \ldots f_n(u))$.

The partition is arbitrary since the equations can be reordered. A new system, w', can now be considered, which is created by duplicating the w system, and the set of variables, v', can be substituted by the corresponding v system:

$$\dot{w}' = h(v, w') \tag{6.10}$$

In this way the w' system is driven by the u system by means of the v variables; w' is called the response system. If all the conditional Lyapunov exponents are less than zero, then the response system will synchronise with the master.

This synchronisation scheme can be developed further. The v subsystem can also be reproduced, creating the v'' subsystem and driving it with the w' variables. The complete slave system is therefore:

$$\begin{aligned} \dot{v}'' &= g(v'', w') \\ \dot{w}' &= h(v, w'') \end{aligned} \tag{6.11}$$

If all the conditional Lyapunov exponents of the (w', v'') systems are negative, then the master (eqn. 6.9) and the slave (eqn. 6.11) will synchronise and, in particular, $v'' \to v$ as $t \to \infty$.

This last set up is called the cascaded synchronisation scheme [3].

An interesting result is the fact that the two systems may be synchronised even in presence of noise or if the driving signal has been altered by a filter [3].

6.1.3 Genetic algorithms

GAs are general-purpose global optimisation techniques based on randomised search and incorporating some aspects of iterative algorithms [4–5].

These algorithms have been inspired by Darwin's theory of evolution: when considering a population which evolves in a particular environment, only the fittest individuals will be able to reproduce, handing down their chromosomes. The descendants of the original population will inherit the qualities that better fit the environment. GAs implement optimisation strategies based on simulation of these natural laws, in order to obtain the fittest individual in the evolutionary sense. Adopting this analogy, the optimal solution corresponds to the fittest individual. GAs search for the optimum starting from a population of points of the function domain (not a single one). This reduces the probability of finding local minima. Moreover, GAs do not require knowledge of the first derivative of the objective function or other auxiliary information. Finally, GAs use probabilistic transition rules during iteration. Adopting a natural analogy, variables involved in optimisation are codified in a particular structure similar to that of a chromosome. For example, a parameter can be translated into a string of l elements (l-bit digits) which will be manipulated by appropriate operators during the evolution of the algorithm. Each string is characterised by a real value, a fitness, strictly connected to the function to be optimised and used to select the more promising elements of the population. The basic string operators that will be applied are:

- Reproduction — consists of duplicating a string.
- Crossover — given two different strings, the operator consists of exchanging substrings defined by some randomly chosen markers.
- Mutation — a variation of a randomly chosen bit, belonging to a selected string.

Reproduction is used to improve the number of fittest individuals in the population, crossover to recombine genetic information between different parents, and mutation to introduce new information into the knowledge base. The strings to which operators apply are chosen according to their fitness. The selection procedure can be implemented by adopting many approaches; the most commonly used are the bias roulette wheel and rank-based selection.

A basic step-by-step genetic algorithm is:

1 Choose at random a fixed number of elements representing the initial population.
2 Evaluate their fitness.
3 Choose the elements of the population according to their actual probabilities.
4 Apply operators, with the respective probabilities, onto chosen elements, obtaining new elements called offspring.
5 Evaluate the fitness of the string obtained.
6 Create a new population using offspring.
7 Go to step 3 until a stop criterion is verified.

6.2 Synchronisation-based identification

In this Section a new approach to chaotic system parameter identification is described, which is based on the Pecora-Carroll cascaded synchronisation approach. The novel identification procedure is fairly general and has been applied to Chua's oscillator as an example because of its above mentioned characteristics. By using the new approach, the chaotic circuit, whose parameters have to be estimated, is considered as a master system and one (at least) of its state variables is used as a driving signal for an identical circuit used as a slave system. The slave system responds to the driving signal following a state trajectory which depends on this input signal and on its parameters that, in general, are different from the master unknown parameters. The distance between the master's state variable, used to drive the slave, and its corresponding slave's state variable, is used to define a performance index.

When the master and the slave have the same parameter set they then synchronise, and their corresponding state variables are asymptotically equal. This means that, in this case, the performance index reaches its global minimum.

6.2.1 Description of the algorithm

In this subsection a more formal description of the new algorithm is given.

Let us consider a nonlinear circuit or system whose mathematical model is supposed to be known, although its parameters are unknown and have to be estimated. Moreover, let us suppose that it is an autonomous circuit or system (this hypothesis is not restrictive, because it is well known that a nonautonomous system can be described by an autonomous model augmenting its original model with suitable additional variables and equations [9–10]). A state representation of this system is considered:

$$\dot{x} = f(x) \tag{6.12}$$

with $x \in \mathcal{R}^n$. As explained in subsection 6.1.2, this system can be divided into two subsystems $n=(v,w)$:

$$\dot{v} = g(v, w)$$
$$\dot{w} = h(v, w) \tag{6.13}$$

where $v = (u_1,...u_m)$, $g = (f_1(u),... f_m(u))$, $w = (u_{m+1},...,u_n)$, $h = (f_{m+1}(u),... f_n(u))$. This partitioned system can be used to realise a cascaded synchronisation scheme as discussed above. Eqns. 6.13 represent the master system, and the slave $x'' = (v'', w')$, driven by:

$$\dot{v}'' = g(v'', w')$$
$$\dot{w}'' = h(v, w') \tag{6.14}$$

This set up is shown in the block diagram of Figure 6.5.

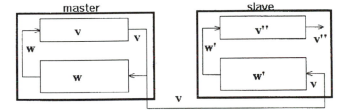

Figure 6.5 The adopted synchronisation scheme

In the slave system, the r unknown parameters $p = (p\_1,..,p\_r)$ can be, at first, arbitrarily assigned.

Let us suppose that v are the only variables available from the master system and that:

$$\{\hat{v}(k)\}, k = 0,1,...,M \tag{6.15}$$

is its corresponding sampled time series (δ being the sampling time). Analogously $\{\hat{v}''(k)\}$ is the time series corresponding to v''.

Let us introduce the following index:

Definition 6.2: Let us define the distance between $\hat{v}$ and $\hat{v}''$ as:

$$I(p) = \sqrt{\sum_{k=0}^{M}\{(\hat{v}_1 - \hat{v}_1''\,(p))^2 + ... + (\hat{v}m - \hat{v}m''\,(p))^2\}} \qquad (6.16)$$

It is clear that this definition is independent of the nature of the time series, so the potential chaotic features are not a problem.

Our identification problem can therefore be formulated as an optimisation problem. In fact, the master and the slave will synchronise if and only if they have identical parameters and, in this case, as from definition 6.2, the index in eqn. 16 has the global minimum. Therefore the parameters p must be changed according to this goal; GAs have been used to achieve this. In particular, a GA in a standard form has been adopted [5]. Besides reproduction, one point crossover and mutation, the elitist strategy has also been utilised.

The time series in eqn. 15 is used to drive a simulated slave system whose parameters are chosen by a GA-based program. It is worth noting that the time discretisation, which is necessary for computer simulation of the slave, inevitably introduces a numerical error that is an increasing function of the discretisation step size. This aspect has to be taken into account in order to evaluate the quality of the obtained results.

6.2.2 Identification of Chua's oscillator

The above procedure is now applied to the case of Chua's oscillator.

The driving signal can be chosen among x, y and z, the proper choice depending on the conditional Lyapunov exponents and so on the considered systems parameters. This means that, in order to obtain synchronisation, some parameter choices require the x variable to be the driving signal and some others require y or z. In the following, the case in which x is used as the driving signal is discussed. However, the procedure is quite general and an example in which the z variable is used to drive the slave systems is presented in the following Section. The state eqns. 1 can be partitioned into two subsystems: the first one, composed pf x only, and the second one, composed of y and z. Using the above terminology, $v = x$ while $w = (y, z)$; then, the slave (driven by x) is described as:

$$\ddot{x}'' = \alpha(y' - l(x''))$$
$$\dot{y}' = x - y' + z'$$
$$\dot{z}' = -\beta y' - \gamma z'$$

(6.17)

where the $r = 5$ parameters are $p = (\alpha, \beta, \gamma, m_0, m_1)$ and the objective index is:

$$I(p) = \sqrt{\sum_{k=0}^{M} \{(\hat{x}_1 - \hat{x}''(p))^2\}}$$

(6.18)

The slave system has been simulated by using a fixed step size fourth-order Runge-Kutta algorithm; this step size has been chosen to be equal to the sampling time, δ, of the driving time series, $\hat{x}$.

6.3 Experimental examples

In this Section three different examples of the application of the new method to Chua's oscillator are reported. The parameters of the Chua's oscillator used as master circuit have been fixed to known values, and the identification procedure has been applied in order to recover these values from only the given time series.

In the first case the parameters have been chosen to obtain a double scroll attractor, that is: $\alpha = 9$, $\beta = 14 \cdot 286$, $\gamma = 0$, $m_0 = -0 \cdot 142857$, $m_1 = 0 \cdot 285714$. The sampling time has been chosen as $\delta = 0 \cdot 1$ and $M = 2000$ samples have been considered. x has been used as the synchronisation signal. The results of the identification are shown in Table 6.1 together with the parameters of the adopted genetic algorithm.

In order to evaluate the obtained results, the attractor generated by the master system and that obtained with the estimated parameters, when they are disconnected, have been overlapped in Figure 6.6; in Figure 6.7 the synchronisation signals x and x'', both with synchronisation error $e = x - x''$, are shown.

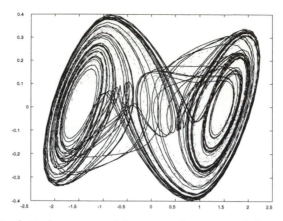

Figure 6.6   *Overlapped attractors of the separated master and slave systems in the x-y plane. The dotted line refers to the master and the solid one is used for the slave*

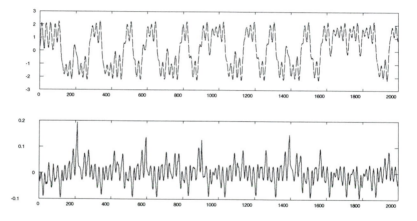

Figure 6.7   *Upper trace: overlapped variables x and x″ of the master and slave systems for the double scroll. The dotted line refers to the master and the solid one is used for the slave. Lower trace: synchronisation error e = x-x″*

In the second example the parameters have been chosen to obtain the attractor shown in Figure 6.3, that is: $\alpha = -4\cdot08585$, $\beta = -2$, $\gamma = 0$, $m_0 = -0\cdot142857$, $m_1 = 0\cdot285714$. In this case the sampling time has been chosen as $\delta = 0\cdot1$ with $M = 2000$ samples and the results are shown in Table 6.1.

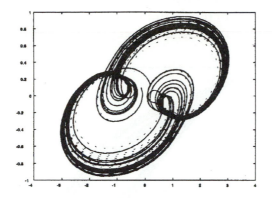

Figure 6.8   *Overlapped attractors of the separated master and slave systems in the x–y plane for attractor no.1. The dotted line refers to the master and the solid one is used for the slave*

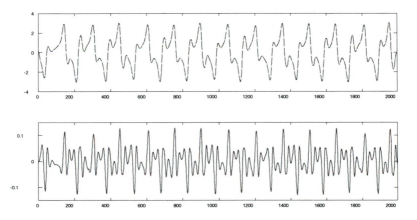

Figure 6.9   *Upper trace: overlapped variables z and z″ of the master and slave systems for attractor no. 1. The dotted line refers to the master and the solid one is used for the slave. Lower trace: synchronisation error e = z–z″*

In this second example synchronisation can only be accomplished by using the z variable as the driving signal. Again the original attractor and the one corresponding to the estimated parameter set have been overlapped as shown in Figure 6.8; in Figure 6.9 the synchronisation signals z and $z″$, both with the synchronisation error $e=z-z″$, are shown.

In the last case the parameters have been chosen to obtain attractor no.2, that is: $\alpha = 6{\cdot}579$, $\beta = 10{\cdot}989$, $\gamma = -0{\cdot}0447$, $m_0 = -0{\cdot}18197$, $m_1 = 0{\cdot}3477$.

The sampling time has been chosen as $\delta = 0{\cdot}01$ and $M = 6000$

samples have been considered. *x* has been used as the synchronisation signal. The results of the identification are shown in Table 6.1 together with the parameters of the adopted genetic algorithm.

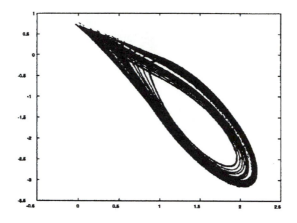

Figure 6.10   *Overlapped attractors of the separated master and slave systems in the x–y plane for attractor no.2. The dotted line refers to the master and the solid one is used for the slave*

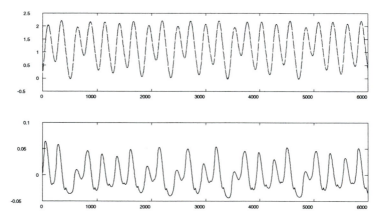

Figure 6.11   *Upper trace: overlapped variables x and x″ of the master and slave systems for attractor no.2. The dotted line refers to the master and the solid one is used for the slave. Lower trace: synchronisation error e = x–x″*

Table 6.1 Parameters and results of the three examples

Example 1		Example 2		Example 3	
Master	Slave	Master	Slave	Master	Slave
$a = 9$	$a = 9.77$	$a = -4.088$	$a = -4.113$	$a = 6.579$	$a = 6.195$
$b = 14.286$	$b = 14.619$	$b = -2$	$b = -2.157$	$b = 10.898$	$b = 10.695$
$g = 0$	$g = 0.157$	$g = 0$	$g = -0.013$	$g = -0.0447$	$g = -0.041$
$m_0 = -0.14285$	$m_0 = -0.122$	$m_0 = -0.1428$	$m_0 = -0.1221$	$m_0 = -0.1197$	$m_0 = -0.17$
$m_1 = 0.285716$	$m_1 = 0.277$	$m_1 = 0.285714$	$m_1 = 1.4922$	$m_1 = 0.3477$	$m_1 = 0.35$
index value		index value		index value	
$I(p) = 1.5721$		$I(p) = 1.4922$		$I(p) = 2.0539$	
GA parameters		GA parameters		GA parameters	
popsize = 100		popsize = 100		popsize = 80	
gen. = 100		gen. = 140		gen. = 200	
p. cros. = 0.6		p. cros. = 0.6		p. cros. = 0.6	
p. mut. = 0.03		p. mut. = 0.006		p. mut. = 0.001	

In order to evaluate the obtained results, the attractor generated by the master system and that obtained with the estimated parameters, when they are disconnected, have been overlapped in Figure 6.10; in Figure 6.11 the synchronisation signals x and x'', both with the synchronisation error $e = x - x''$, are shown.

The accuracy of the estimation process can be increased if more samples, smaller step sizes and more generations are used. Of course this implies increased computational costs so a trade off is necessary. Moreover, some dynamics are more difficult to estimate, that is, the parameters of the master must be very close to those of the slave in order to obtain the correct synchronisation. This feature is related to the so-called structural stability [10] of the considered dynamic, namely to the feature of the system which retains its qualitative properties under small perturbations of the parameters or of the model.

6.4 Conclusions

In this Chapter a new method of identifying the parameters of nonlinear circuits has been presented, based on the concepts of synchronisation of nonlinear circuits. The new procedure has been formulated as a global optimisation problem and it has been solved by using a genetic algorithm.

The method has been applied to the estimation of the five dimensionless parameters of the chaotic Chua's oscillator and three experimental examples have been reported. The accuracy of the method has also been discussed.

The advantages of the introduced algorithm are numerous; among these is its intrinsic low sensitivity to noise due to the robustness of the synchronisation framework [3].

With the proposed approach a circuit model for chaotic behaviour could be obtained. In fact, many different attractors have been observed in nonlinear circuits and the introduced strategy represents a useful tool for determining the parameters of a circuit model which best fit a chaotic time series [11]. Moreover, it could be used to estimate the parameters of the nonlinear circuit used as a modulator in a chaotic carrier cryptography system.

6.5 References

1   Pecora, L. M. and Carroll, T. L.: 'Synchronization in chaotic systems', Phys. Rev. Lett. **64**, pp. 821–824, 1990

2   Hasler, M.: 'Synchronization principles and applications', ISCAS'94 tutorials, pp.314–327, 1994

3   Carroll, T. L.: 'Communicating with use of filtered, synchronized, chaotic signals', *IEEE Trans. Circuits Syst. Fundam. Theory Appl.*, **42**, (3) pp. 105–110, 1995

4   Cuomo, K. M., Oppenheim A. V., and Strogatz, S. H.: 'Synchronization of Lorenz-based chaotic circuits with applications to communications', *IEEE Trans. Circuits Syst. Analog Digit. Signal Process. II*, **40**, (10), pp.626–633, 1993

5   Goldberg, D. E.: *Genetic algorithms in search, optimization and machine learning* (Addison Wesley, 1989)

6   Caponetto, R., Fortuna, L., Graziani, S., and Xibilia, M. G.: 'Genetic algorithms and applications in system engineering: a survey', *IEEE Trans. Instrum. Meas.* **15** (3), pp. 143–156, 1993

7   Madan, R.: *Chua's circuit: A paradigm for chaos*, (World scientific pub. co., Singapore, 1993)

8   Chua, L. O.: 'Global unfolding of Chua's circuit', *IEICE Trans. Fundam. Electron. Commun. Comput. Sci.* **E76-A**, pp.704–734, 1993

9   Parker, T. S., Chua, L. O.: 'Chaos: a tutorial for engineering', *Proc. IEEE*, **75**, (8), pp. 982-1008, 1987

10 Guckenheimer, J., and Holmes, P.: *Nonlinear oscillations, dynamical systems, and bifurcations of vector fields* (Springer-Verlag, 1983)

11 Baglio, S., Cristaudo, R., Fortuna, L., and Manganaro, G.: 'Complexity in an industrial flyback converter', *J. Circuits, Sys. Comput.*, 1995

Chapter 7
Job shop scheduling

T. Yamada and R. Nakano

7.1 Introduction

Scheduling is the allocation of shared resources over time to competing activities, and has been the subject of a significant amount of literature in the operations research field. Emphasis has been on investigating machine scheduling problems where jobs represent activities and machines represent resources; each machine can process at most one job at a time.

Table 7.1   A 3×3 problem

job	operations routing (processing time)		
1	1 (3)	2 (3)	3 (3)
2	1 (2)	3 (3)	2 (4)
3	2 (3)	1 (2)	3 (1)

The $n \times m$ minimum makespan general job shop scheduling problem, hereafter referred to as the JSSP, can be described by a set of n jobs $\{J_i\}_{1 \leq i \leq n}$ which is to be processed on a set of m machines $\{M_r\}_{1 \leq r \leq m}$. Each job has a technological sequence of machines to be processed. The processing of job J_j on machine M_r is called the operation O_{jr}. Operation O_{jr} requires the exclusive use of M_r for an uninterrupted duration p_{jr}, its processing time. A schedule is a set of completion times for each operation $\{c_{jr}\}_{1 \leq j \leq n, 1 \leq r \leq m}$ which satisfies those constraints. The time required to complete all the jobs is called the makespan L. The objective when solving or optimising this general problem is to determine the schedule which minimises L. An example of a 3 × 3 JSSP is given in Table 7.1. The data includes the routing of each job through each machine and the processing time for each operation (in parentheses).

The Gantt chart is a convenient way of visually representing a solution of the JSSP. An example of a solution for the 3×3 problem in Table 7.1 is given in Figure 7.1.

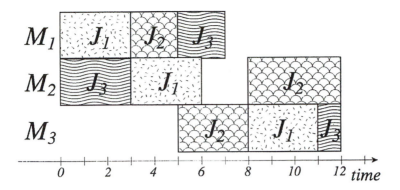

Figure 7.1   A Gantt chart representation of a solution for a 3 × 3 problem

The JSSP is not only $\mathcal{NP}$-hard, but it is one of the worst members in the class. An indication of this is given by the fact that one 10×10 problem formulated by Muth and Thompson [18] remained unsolved for over 20 years.

Besides exhaustive search algorithms based on branch and bound methods, several approximation algorithms have been developed. The most popular ones in practice are based on priority rules and active schedule generation [21]. A more sophisticated method called shifting bottleneck (SB) has been shown to be very successful [1]. Additionally, stochastic approaches such as simulated annealing (SA), tabu search [11,33] and genetic algorithms (GAs) have been recently applied with good success.

This Chapter reviews a variety of GA applications to the JSSP. We begin our discussion by formulating the JSSP by a disjunctive graph. We then look at domain independent binary and permutation representations, followed by an active schedule representation with GT crossover and the genetic enumeration method. Section 7.7 discusses a method for integrating local optimisation directly into GAs. Section 7.8 discusses performance comparison using the well known Muth and Thompson benchmark and the more difficult ten tough problems.

7.2 Disjunctive graph

The JSSP can be formally described by a disjunctive graph $G = (V, C \cup D)$, where:

- *V* is a set of nodes representing operations of the jobs together with two special nodes, a source (0) and a sink *, representing the beginning and end of the schedule, respectively.
- *C* is a set of conjunctive arcs representing technological sequences of the operations.
- *D* is a set of disjunctive arcs representing pairs of operations which must be performed on the same machines.

The processing time for each operation is the weighted value attached to the corresponding nodes. Figure 7.2 shows this in a graph representation for the problem given in Table 7.1.

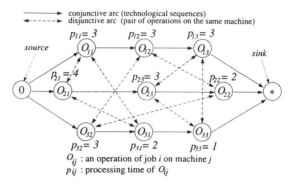

O_{ij} : an operation of job i on machine j
p_{ij} : processing time of O_{ij}

Figure 7.2   *A disjunctive graph of a 3 × 3 problem*
conjunctive arc (technological sequences)
disjunctive arc (pair of operations on the same machine)
O_{ij}: an operation of job i on machine j
p_{ij}: processing time of O_{ij}

Job shop scheduling can also be viewed as defining the ordering between all operations that must be processed on the same machine, i.e. to fix precedences between these operations. In the disjunctive graph model, this is done by turning all undirected (disjunctive) arcs into directed ones. A selection is a set of directed arcs selected from disjunctive arcs. By definition, a selection is complete if all the disjunctions are selected. It is consistent if the resulting directed graph is acyclic.

A schedule uniquely obtained from a consistent complete selection by sequencing operations as early as possible is called a semiactive schedule. In a semiactive schedule, no operation can be started earlier without altering the machining sequences. A consistent complete selection and the corresponding semiactive schedule can be represented by the same symbol *S* without confusion. The makespan *L* is given by the length of the longest weighted path from source to sink in this graph. This path, $\mathscr{P}$, is called a critical path and is composed of a sequence of critical operations. A sequence of consecutive critical operations on the same machine is called a critical block.

The distance between two schedules S and T can be measured by the number of differences in the processing order of operations on each machine [19]. In other words, it can be calculated by summing those disjunctive arcs that have directions which are different between S and T. We call this distance the disjunctive graph (DG) distance. Figure 7.3 shows the DG distance between two schedules. The two disjunctive arcs drawn by thick lines in schedule (b) have directions which differ from those of schedule (a), and therefore the DG distance between (a) and (b) is 2.

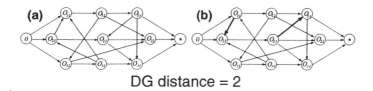

DG distance = 2

Figure 7.3   The DG distance between two schedules

7.2.1 Active schedules

The makespan of a semiactive schedule may often be reduced by shifting an operation to the left without delaying other jobs. Such reassigning is called a permissible left shift and a schedule with no more permissible left shifts is called an active schedule. An optimal schedule is clearly active so it is safe and efficient to limit the search space to the set of all active schedules. An active schedule is generated by the GT algorithm proposed by Giffler and Thompson [13], which is described in algorithm 7.2.1. In the algorithm, the earliest starting time ES(O) and earliest completion time EC(O) of an operation O denote its starting and completion times when processed with the highest priority amongst all currently schedulable operations on the same machine. An active schedule is obtained by repeating the algorithm until all operations are processed. In Step 3, if all possible choices are considered, all active schedules will be generated, but the total number will still be very large.

Figure 7.4 shows how the GT algorithm works. In the figure, O_{11} is identified as O_{jr} and M_1 as M_r. Then, O_{31} is selected from the conflict set and scheduled. After that, the conflict set and earliest starting times of operations are updated.

7.3 Binary representation

As described in the previous Section, a (semiactive) schedule is obtained by turning all undirected disjunctive arcs into directed ones. Therefore, by labelling each directed disjunctive arc of a schedule as 0 or 1 according to its direction, a schedule can be represented by a binary string of length $mn(n-1)/2$. Figure 7.5 shows a labelling example, where an arc connecting O_{ij} and O_{kj} ($i < k$) is labelled as 1 if the arc is directed from O_{ij} to O_{kj} (so O_{ij} is processed prior to O_{kj}) or 0, otherwise. It should be noted that the DG distance between schedules and the Hamming distance between the corresponding binary strings can be identified through this binary mapping.

Algorithm 7.2.1   *GT algorithm*

---

1 Let D be a set of all the earliest operations in a technological sequence not yet scheduled and O_{jr} be an operation with the minimum EC in D: $O_{jr} = \arg\min\{O \in D \mid EC(O)\}$.
2 Assume $i-1$ operations have been scheduled on M_r. A *conflict set* $C[M_r,i]$ is defined as: $C[M_r,i] = \{O_{kr} \in D \mid O_{kr}$ on M_r, $ES(O_{kr}) < EC(O_{jr})\}$.
3 Select an operation $O \in C[M_r,i]$.
4 Schedule O as the ith operation on M_r with its completion time equal to $EC(O)$.

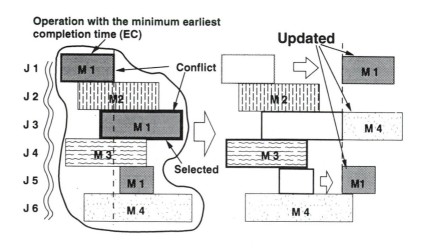

Figure 7.4   *Giffler and Thompson's active schedule generation*

A conventional GA using this binary representation was proposed by Nakano and Yamada [19]. An advantage of this approach is that conventional genetic operators, such as one-point, two-point and uniform crossovers can be applied without any modification. However, a resulting new bit string generated by crossover may not represent a schedule, and such a bit string would be called illegal. There are two approaches for this problem: one is to repair an illegal string and the other is to impose a penalty for the illegality. The following Sections will elaborate on one example of the former approach.

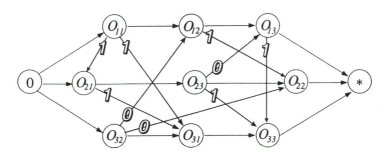

Figure 7.5   Labelling disjunctive arcs

7.3.1 Local harmonisation

A repairing procedure that generates a feasible bit string, as similar to an illegal one as possible, is called the harmonisation algorithm [19]. The Hamming distance is used to assess the similarity between two bit strings. The harmonisation algorithm goes through two phases: local harmonisation and global harmonisation. The former removes the ordering inconsistencies within each machine, and the latter removes the ordering inconsistencies between machines.

Local harmonisation works separately for each machine: thus, the following merely explains how it works for one machine. Here we are given an original illegal bit string. The bit string indicates the processing priority on the machine and may include an ordering inconsistency within the machine, for example, job 1 must be prior to job 2, job 2 must be prior to job 3, but job 3 must be prior to job 1. The local harmonisation can eliminate such a local inconsistency. At first, the algorithm regards the operation having the highest priority as the one to process first. When there is more than one candidate, it selects one of them. Then it removes the priority inconsistencies relevant to the top operation. By repeating the above, the local inconsistency can be completely removed. Local harmonisation goes halfway towards generating a feasible bit string.

7.3.2 Global harmonisation

Global harmonisation removes ordering inconsistencies between machines. It is embedded in a simple scheduling algorithm. First, the scheduling algorithm is explained. Given the processing priority generated by the above local harmonisation as well as the technological sequences and processing time for each operation, the scheduling algorithm polls jobs, checking if any job can be scheduled, and schedules an operation of a job that can be scheduled. It stops if no more jobs can be scheduled due to a global inconsistency, i.e. a deadlock happens. Global harmonisation is called whenever such a deadlock occurs.

The algorithm works as follows. For each job, j, the algorithm considers *next* (j), the job j operation to be scheduled next, and *next* (j).*machine*, the machine which processes $next(j)$. The algorithm calculates how far in the processing priority it is from $next(next(j).machine)$, the next operation on the machine, to $next(j)$. The algorithm selects the job with the minimum distance. When there is more than one candidate, it selects one of them. Then it removes the priority inconsistencies relevant to the permutation, and returns control to the scheduling algorithm.

Thus the scheduling algorithm generates a feasible bit string in cooperation with the global harmonisation. It is not always guaranteed that the above harmonisation will generate a feasible bit string closest to the original illegal one, but the resulting one will be reasonably close and the harmonisation algorithms are quite efficient.

7.3.3 Forcing

An illegal bit string produced by genetic operations can be considered as a genotype, and a feasible bit string generated by any repairing method can be regarded as a phenotype. Then the former is an inherited character and the latter is an acquired one. Note that the repairing stated above is only used for fitness evaluation of the original bit string; that is, repairing does not mean the replacement of bit strings.

Forcing means the replacement of the original string with a feasible one. Hence, forcing can be considered as the inheritance of an acquired character, although it is not widely believed that such inheritance occurs in nature. Since frequent forcing may destroy whatever potential and diversity of the population, it is limited to a small number of elites. Such limited forcing brings about at least two merits: a significant improvement in the convergence speed and the solution quality. Experiments have shown how it works [19].

7.4 Permutation representation

As described in Section 7.2, the JSSP can be viewed as an ordering problem just like the travelling salesman problem (TSP). For example, a schedule can be represented by the set of permutations of jobs on each machine, in other words, m partitioned permutations of operation numbers, which is called a job sequence matrix. Figure 7.6 shows a job sequence matrix of the same solution as that given in Figure 7.1. The advantage of this representation is that the GA operators used to solve the TSP can be applied without further modifications, because each job sequence is equivalent to path representation in the TSP.

$$M_1 \qquad M_2 \qquad M_3$$
$$1\ 2\ 3 \qquad 3\ 1\ 2 \qquad 2\ 1\ 3$$

Figure 7.6   A job sequence matrix for a 3 × 3 problem

7.4.1 Subsequence exchange crossover

A crossover operator called the subsequence exchange crossover (SXX) was proposed by Kobayashi, Ono and Yamamura [15]. The SXX is a natural extension of the subtour exchange crossover for TSPs presented by the same authors [14]. Let two job sequence matrices be p_0 and p_1. A pair of subsequences, one from p_0 and the other from p_1 on the same machine, is called exchangeable if and only if the two halves consist of the same set of jobs. The SXX searches for exchangeable subsequence pairs in p_0 and p_1 on each machine and interchanges each pair to produce new job sequence matrices k_0 and k_1. Figure 7.7 shows an example of the SXX for a 6 × 3 problem.

	M_1	M_2	M_3
p_0	123456	321564	235614
p_1	621345	326451	635421
		▼	
k_0	213456	325164	263514
k_1	612345	326415	356421

Figure 7.7   Subsequence exchange crossover (SXX)

If all jobs in a job subsequence s_0 in p_0 on a machine are positioned consecutively in s_1 in p_1, s_0 and s_1 are exchangeable. By checking for all s_0 in p_0 systematically, if there exists a corresponding s_1 in p_1, all of the exchangeable subsequence pairs in p_0 and p_1 on the machine can be enumerated in $O(n^2)$ [29], so the SXX requires a computational complexity of $O(mn^2)$.

Although a job sequence matrix obtained from the SXX always represents valid job permutations, it does not necessarily represent a schedule. To obtain a schedule from illegal offspring, some repairing mechanism such as the global harmonisation described in Section 7.3 is also required. Instead of using global harmonisation, the GT algorithm is used as a repairing mechanism, together with the described forcing, to modify any job sequence matrix into an active schedule. A small number of swap operations designated by the GT algorithm are applied to repair job sequence matrices.

7.4.2 Permutation with repetition

Instead of using an m-partitioned permutation of operation numbers, like the job sequence matrix defined in the previous subsection, another representation which uses an unpartitioned permutation with m repetitions of job numbers was employed by Bierwirth [6]. In this permutation, each job number occurs m times. By scanning the permutation from left to right the kth occurrence of a job number refers to the kth operation in the technological sequence of this job (see Figure 7.8). In this representation, it is possible to avoid schedule operations whose technological predecessors have not been scheduled yet. Therefore, any individual is decoded to a schedule, but two or more different individuals can be decoded to an identical schedule.

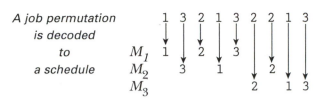

Figure 7.8   *A job sequence (permutation with repetition) for a 3 × 3 problem is decoded to a schedule, which is equivalent to the one in Figure 7.1*

The well used order crossover and partially mapped crossover for TSP are naturally extended for this representation (they are called the generalised order crossover (GOX) and generalised partially mapped crossover (GPMX)). A new precedence preservative crossover (PPX) is also proposed in [7]. The PPX perfectly respects the absolute order of genes in parental chromosomes. A template bit string h of length mn is used to define the order in which genes are drawn from p_0 and p_1. A gene is drawn from one parent and it is appended to the offspring chromosome. The corresponding gene is deleted in the other parent (see Figure 7.9). This step is repeated until both parent chromosomes are empty and the offspring contains all genes involved. The idea of forcing described in Section 7.3 is combined with the permissible left shift described in Section 7.2.1; new chromosomes are modified to active schedules by applying permissible left shifts.

$$
\begin{array}{ll}
p_0 & (3\ 2)\ 2\ 2\ 3\ 1\ 1\ 1\ 3 \\
h & (0\ 0)\ 1\ 1\ 1\ 1\ 0\ 0\ 0 \\
p_1 & 1\ 1\ 3\ 2\ 2\ 1\ 2\ 3\ 3 \\
k & (3\ 2)\ 1\ 1\ 2\ 1\ 2\ 3\ 3
\end{array}
$$

Figure 7.9   *Precedence preservative crossover (PPX)*

7.5 Heuristic crossover

The earlier Sections were devoted to representing solutions in generic forms such as bit strings or permutations so that conventional crossover operators could be applied without further modifications. Because of the complicated constraints of the problem, however, individuals generated by a crossover operator are often infeasible and require several steps of a repairing mechanism. The following properties are common to these approaches:

Algorithm 7.5.1	GT crossover

1 Same as Step 1 of algorithm 7.2.1.
2 Same as Step 2 of algorithm 7.2.1.
3 Select one of the parent schedules $\{p_0, p_1\}$ according to the value of H_{ir} as $p = p_{Hir}$. Select an operation $O \in C[M_r, i]$ that has been scheduled in p earliest among $C[M_r, i]$.
4 Same as Step 4 of algorithm 7.2.1.

- Crossover operators are problem independent and they are separated from schedule builders.
- An individual does not represent a schedule itself but its gene codes give a series of decisions for a schedule builder to generate a schedule.

Obviously one of the advantages of the GA is its robustness over a wide range of problems with no requirement for domain specific adaptations. Therefore, the crossover operators should be domain independent and separated from domain specific schedule builders. However, from the viewpoint of performance, it is often more efficient to directly incorporate domain knowledge into the algorithm to skip wasteful intermediate decoding steps. Thus the GT crossover proposed by Yamada and Nakano [30] has the following properties instead:

- The GT crossover is a problem dependent crossover operator that directly utilises the GT algorithm. In the crossover, parents cooperatively give a series of decisions to the algorithm to build new offspring, namely active schedules.
- An individual represents an active schedule, so there is no repairing scheme required.

7.5.1 GT crossover

Let H be a binary matrix of size $n \times m$ [30, 10]. Here $H_{ir} = 0$ means that the ith operation on machine r should be determined by using the first parent and $H_{ir} = 1$ by the second parent. The role of H_{ir} is similar to that of h described in Section 7.4.2. Let the parent schedules be p_0 and p_1 as always. The GT crossover can be defined by modifying Step 3 of algorithm 7.2.1 as shown in algorithm 7.5.1. It tries to reflect the processing order of the parent schedules to their offspring. It should be noted that if the parents are identical to each other, the resulting new schedule is also identical to the parents. In general, the new schedule inherits partial job sequences of both parents in different ratios depending on the number of 0s and 1s contained in H.

The GT crossover generates only one schedule at once. Another schedule is generated by using the same H but changing the roles of p_0 and p_1. Thus, two new schedules are generated that complement each other. The outline of the GT crossover is described in Figure 7.10.

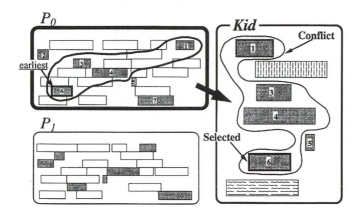

Figure 7.10   GT crossover

Mutation can be put into algorithm 7.5.1 by occasionally selecting the nth $(n > 1)$ earliest operation in $C[M_r, i]$ with a low probability inversely proportional to n in Step 3 of algorithm 7.5.1.

7.6 Genetic enumeration

A method for using the bit string representation and simple crossover used in simple GAs, and at the same time incorporating problem specific heuristics, was proposed by Dorndorf and Pesch [12, 22]. They interpret an individual solution as a sequence of decision rules for domain specific heuristics such as the GT algorithm and the shifting bottleneck procedure.

7.6.1 Priority rule based GA

Priority rules are the most popular and the simplest heuristics for solving the JSSP. They are rules used in Step 3 of algorithm 7.2.1 to resolve a conflict by selecting an operation O from the conflict set $C[M_r, i]$. For example, a priority rule called SOT rule (shortest operation time rule) selects the operation with the shortest processing time from the conflict set. Twelve such simple rules are used in [12, 22] including SOT rule, LRPT rule (longest remaining processing time rule) and FCFS rule (first come first serve rule) such that they are partially complementary in order to select each member in the conflict set.

Each individual of the priority rule based GA (P-GA) is a string of length $mn-1$, where the entry in the ith position represents one of the 12 priority rules used to resolve the conflict in the ith iteration of the GT algorithm. A simple crossover that exchanges the substrings of two cut strings is applied.

Algorithm 7.6.1 The shifting bottleneck procedure (SB I)

---

1 Set $S = \varnothing$ and make all machines unsequenced.
2 Solve a one-machine scheduling problem for each unsequenced machine.
3 Among the machines considered in Step 2, find the bottleneck machine and add its schedule to S. Make the machine sequenced.
4 Reoptimise all sequenced machines in S.
5 Go to Step 3 unless S is completed; otherwise stop.

7.6.2 Shifting bottleneck based GA

The shifting bottleneck (SB) proposed by Adams *et al.* [1] is a powerful heuristic for solving the JSSP. In this method, a one-machine scheduling problem (a relaxation of the original JSSP) is solved for each machine not yet sequenced, and the outcome is used to find a bottleneck machine, i.e. a machine having the longest makespan. Every time a new machine has been sequenced, the sequence of each previously sequenced machine is subject to reoptimisation. The SB consists of two subroutines: the first one (SB I) repeatedly solves one-machine scheduling problems; the second one (SB II) builds a partial enumeration tree where each path from the root to a leaf is similar to an application of SB I. The outline of the SB I is described in algorithm 7.6.1. Please refer to [1, 2, 33] as well as [12, 22] for more details.

Besides using the genetic algorithm as a metastrategy to optimally control the use of priority rules, another genetic algorithm described in Dorndorf and Pesch [12, 22] controls the selection of nodes in the enumeration tree of the shifting bottleneck heuristic; it is called the shifting bottleneck based genetic algorithm (SB-GA). Here, an

individual is represented by a permutation of machine numbers
$1 \ldots m$, where the entry in the ith position represents the machine
selected in Step 3 in place of a bottleneck machine in the ith iteration
of algorithm 7.6.1. A cycle crossover operator is used as the crossover
for this permutation representation.

7.7 Genetic local search

It is well known that GAs can be enhanced by incorporating local
search methods, such as neighborhood search, into them. The result
of such an incorporation is often called genetic local search (GLS)
[26]. In this framework, an offspring obtained by a recombination
operator, such as crossover, is not included in the next generation
directly but is used as a seed for the subsequent local search. The local
search moves the offspring from its initial point to the nearest locally
optimal point, which is included in the next generation.

    This Section briefly reviews the basics of neighbourhood search,
neighbourhood structures for the JSSP and an approach to
incorporating a local neighbourhood search into a GA to solve the
problems.

7.7.1 Neighbourhood search

Neighbourhood search is a widely used local search technique to solve
combinatorial optimisation problems. A solution x is represented as a
point in the search space, and a set of solutions associated with x is
defined as neighbourhood $N(x)$. $N(x)$ is a set of feasible solutions reach-
able from x by exactly one transition, i.e. a single perturbation of x.

    An outline of a neighbourhood search for minimising $V(x)$ is
described in algorithm 7.7.1, where x denotes a point in the search
space and $V(x)$ denotes its evaluation value.

Algorithm 7.7.1 Neighbourhood search

- Select a starting point: $x = x_0 = x_{best}$.

do   1 Select a point $y \in N(x)$ according to the given criterion based
on the value $V(y)$. Set $x = y$.
       2 If $V(x) < V(x_{best})$ then set $x_{best} = x$.

until some termination condition is satisfied.

The criterion used in Step 1 in algorithm 7.7.1 is called the choice criterion, by which the neighbourhood search can be categorised. For example, a descent method selects a point $y \in N(x)$ such that $V(y) < V(x)$. A stochastic method probabilistically selects a point according to the metropolis criterion, i.e. $y \in N(x)$ is selected with probability 1 if $V(y) < V(x)$; otherwise, with probability:

$$P(y) = exp(-\Delta V/T), \text{ where } \Delta V = V(y) - V(x) \tag{7.1}$$

Here, P is called the acceptance probability. Simulated annealing (SA) is a method in which parameter T (called the temperature) decreases to zero following an annealing schedule as the iteration step increases.

Algorithm 7.7.2 Multistep crossover fusion (MSXF)

- Let p_0, p_1 be parent solutions.
- Set $x = p_0 = q$.
do   • For each member $y_i \in N(x)$, calculate $d(y_i, p_1)$.
   • Sort $y_i \in N(x)$ in ascending order of $d(y_i, p_1)$.
   do 1 Select y_i from $N(x)$ randomly, but with a bias in favor of y_i with a small index i.
      2 Calculate $V(y_i)$ if y_i has not yet been visited.
      3 Accept y_i with probability one if $V(y_i) \leq V(x)$, and with $P_c(y_i)$ otherwise.
      4 Change the index of y_i from i to n, and the indices of y_k ($k \in \{i+1, i+2,...,n\}$) from k to $k - 1$.
   until y_i is accepted.
   • Set $x = y_i$.
   • If $V(x) < V(q)$ then set $q = x$.
until some termination condition is satisfied.
- q is used for the next generation.

7.7.2 Multistep crossover fusion

Reeves has been exploring the possibility of integrating local optimisation directly into a simple GA with bit string representations and has proposed the neighbourhood search crossover (NSX) [23]. Let any two individuals be x and z. An individual y is called intermediate between x and z, written as $x \Diamond y \Diamond z$, if and only if $d(x, z) = d(x, y) + d(y, z)$ holds, where x, y and z are represented in binary strings and $d(x, y)$ is the Hamming distance between x and y. Then the

kth-order two neighbourhood of x and z is defined as the set of all intermediate individuals at a Hamming distance of k from either x or z. Formally,

$$N_k(x, z) = \{y \mid x \Diamond y \Diamond z \text{ and } (d(x, y) = k \text{ or } d(y, z) = k)\}$$

Given two parent bit strings, p_0 and p_1, the neighbourhood search crossover of order k (NSX$_k$) will examine all individuals in $N_k(p_0, p_1)$, and pick the best as the new offspring.

Yamada and Nakano extended the idea of the NSX to make it applicable to more complicated problems such as job shop scheduling and proposed the multistep crossover fusion (MSXF): a new crossover operator with a built in local search functionality [31, 32, 34]. The MSXF has the following characteristics compared to the NSX.

- It can handle more generalised representations and neighbourhood structures.
- It is based on a stochastic local search algorithm.
- Instead of restricting the neighbourhood by a condition of intermediateness, a biased stochastic replacement is used.

A stochastic local search algorithm is used for the base algorithm of the MSXF. Although the SA is a well known stochastic method and has been successfully applied to many problems as well as to the JSSP, it would be unrealistic to apply the full SA to suit our purpose because it would consume too much time by being run many times in a GA run. A restricted method with a fixed temperature parameter $T = c$ might be a good alternative. Accordingly, the acceptance probability used in algorithm 7.7.1 is rewritten as:

$$P_c(y) = exp\left(-\frac{\Delta V}{c}\right), \Delta V = V(y) - V(x), \ c : \text{const} \qquad (7.2)$$

Let the parent schedules be p_0 and p_1, and let the distance between any two individuals x and y in any representation be $d(x, y)$. If x and y are schedules, then $d(x, y)$ is the DG distance. Crossover functionality can be incorporated into algorithm 7.7.1 by setting $x_0 = p_0$ and adding a greater acceptance bias in favour of $y \in N(x)$ having a small $d(y, p_1)$. The acceptance bias in the MSXF is controlled by sorting $N(x)$ members in ascending order of $d(y_i, p_1)$ so that y_i with a smaller index i has a smaller distance $d(y_i, p_1)$. Here $d(y_i, p_1)$ can be estimated easily if $d(x, p_1)$ and the direction of the transition from x to y_i are known; it is not necessary to generate and evaluate y_i. Then y_i is selected from $N(x)$ randomly, but with a bias in favour of y_i with a small index, i. The outline of the MSXF is described in algorithm 7.7.2.

In place of $d(y_i, p_1)$, one can also use $sign(d(y_i, p_1) - d(x, p_1)) + r_e$ to sort $N(x)$ members in algorithm 7.7.2. Here $sign(x)$ denotes the sign of x: $sign(x) = 1$ if $x > 0$, $sign(x) = 0$ if $x = 0$, $sign(x) = -1$ otherwise. A small random fraction r_e is added to randomise the order of members with the same sign. The termination condition can be given, for example, as the fixed number of iterations in the outer loop.

The MSXF is not applicable if the distance between p_0 and p_1 is too small compared to the number of iterations. In such a case, a mutation operator called the multistep mutation fusion (MSMF) is applied instead. The MSMF can be defined in the same manner as the MSXF is except for one point: the bias is reversed, i.e. sort the $N(x)$ members in descending order of $d(y_i, p_1)$ in algorithm 7.7.2.

7.7.3 Neighbourhood structures for the JSSP

For the JSSP, a neighbourhood $N(S)$ of a schedule S can be defined as a set of schedules which can be reached from S by exactly one transition (a single perturbation of S).

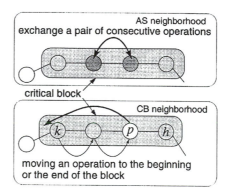

Figure 7.11   Permutation of operations on a critical block

As shown in Section 7.3, a set of solutions of the JSSP can be mapped to a space of bit strings by marking each disjunctive arc as 1 or 0 according to its direction. The DG distance and the Hamming distance in the mapped space are equivalent, and the neighbourhood of a schedule S is a set of all (possibly infeasible) schedules whose DG distances from S are exactly one. Neighbourhood search using this binary neighbourhood is simple and straightforward but not very efficient.

More efficient methods can be obtained by introducing a transition operator that exchanges a pair of consecutive operations only on the critical path and forms a neighbourhood [16, 25]. The transition operator was originally defined by Balas in his branch and bound approach [4]. We call this the adjacent swapping (AS) neighbourhood. DG distances between a schedule and members of its AS neighbourhood are always one, so the AS neighbourhood can be considered as a subset of the bit string neighbourhood.

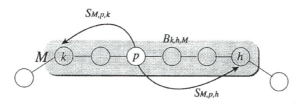

Figure 7.12 $S_{M,p,k}$ and $S_{M,p,h}$ *generation*

Another very powerful transition operator was used in [9, 11]. The transition operator permutes the order of operations in a critical block by moving an operation to the beginning or end of the critical block, thus forming a CB neighbourhood.

A schedule obtained from S by moving an operation within a block to the front of the block is called a before candidate, and a schedule obtained by moving an operation to the rear of the block is called an after candidate. A set of all before and after candidates $N'^c(S)$ may contain infeasible schedules. The CB neighbourhood is given as:

$$N^c(S) = \{S' \in N'^c (S) \mid S' \text{ is a feasible schedule}\}$$

It has been experimentally shown by [35] that the CB neighbourhood is more powerful than the former one.

Active CB neighbourhood

As explained above, before or after candidates are not necessarily executable. In the following, a new neighbourhood similar to the CB neighbourhood is used, each element of which is not only executable, but also active and close to the original. Let S be an active schedule and $B_{k,h,M}$ be a critical block of S on a machine M, where the front and the rear operations of $B_{k,h,M}$ are the kth and the hth operations on M, respectively. Let $O_{p,M}$ be an operation in $B_{k,h,M}$ that is the pth operation on M. Algorithm 7.7.3 generates an active schedule $S_{M,p,k}$ (or $S_{M,p,h}$) by modifying S such that $O_{p,M}$ is moved to the position as close to the front position k (or the rear position h) of $B_{k,h,M}$ as possible. Parts of the algorithm are due to [11]. The new active CB neighbourhood $AN^c(S)$ is now defined as a set of all $S_{M,p,k}$ and $S_{M,p,h}$ over all critical blocks:

$$AN^c(S) = \bigcup_{B_{k,h,M}} \{S' \in \{S_{M,p,k}\}_{k<p<h} \cup \{S_{M,p,h}\}_{k<p<h}, S' \neq S\}$$

7.7.4 Scheduling in the reversed order

Algorithm 7.2.1 and all its variations determine the job sequences from left to right in temporal order. This is because active schedules are defined to have no extra idle periods of machines prior to their operations. However, the idea described below enables the same algorithms to determine the job sequences from right to left with only small modifications.

In general, a given problem of the JSSP can be converted to another problem by reversing all of the technological sequences. The new problem is equivalent to the original one in the sense that reversing the job sequences of any schedule for the original problem results in a schedule for the reversed problem with the same critical path and makespan. It can be seen, however, that an active schedule for the original problem may not necessarily be active in the reversed problem: the active quality is not necessarily preserved.

Algorithm 7.7.3 Modified GT algorithm generating $S_{M,p,k}$ or $S_{M,p,h}$

---

1 Same as Step 1 of algorithm 7.2.1.
2 Same as Step 2 of algorithm 7.2.1.
3 Do **CASE 1** (or **CASE 2**) to generate $S_{M,p,k}$ (or $S_{M,p,h}$)
   CASE 1: $S_{M,p,k}$ generation
   - If $k \leq i \leq p$ and $O_{p,M} \in C[M_r i]$, then set $O = O_{p,M}$.
   - *Otherwise*, select an operation $O \in C[M_r, i]$ that has been scheduled in S earliest among $C[M_r, i]$.
   CASE 2: $S_{M,p,h}$ generation
   - If $i = h$ or $C[M_r, i] = \{O_{p,M}\}$, then set $O = O_{p,M}$.
   - *Otherwise*, select an operation $O \in C[M_r i] \setminus O_{p,M}$ that has been scheduled in S earliest among $C[M_r i] \setminus O_{p,M}$.
4 Same as Step 4 of algorithm 7.2.1.

---

job	routing
1	2(3) 1(4)
2	2(2) 1(4)

Figure 7.13   A simple 2 × 2 problem

For example, the simple 2 × 2 problem described in Figure 7.13 is considered. Figure 7.14(1) shows a solution of this problem, which is active and no more left shifts can improve its makespan. Figure 7.14(2), obtained by reversing Figure 7.14(1), is not active and can be improved by a left shift which moves job 1 prior to job 2 on machine 2, resulting in Figure 7.14(3). Finally Figure 7.14(4) is obtained by reversing Figure 7.14(3) again, which is optimal. As things turn out, Figure 7.14(1) is improved by moving job 1 posterior to job 2 on machine 2, resulting in Figure 7.14(4).

Although repairing a semiactive schedule to an active state improves the makespan, it can be seen from the example above that there sometimes are obvious improvements that cannot be attained only by left shifts. We call a schedule left active if it is an active schedule for the original problem and right active if it is such for the reversed problem. It sometimes happens that a reserved problem is easier to solve than the original. Searching only in the set of left (or right) active schedules may bias the search towards the wrong direction and result in poor local minima. Therefore, left active schedules as well as right active ones should be taken into account together in the same algorithm. In most local search methods, many schedules are generated in a single run; therefore, it would be better to apply this reversing and repairing method periodically to change the scheduling directions rather than to reverse and repair every schedule each time it is generated.

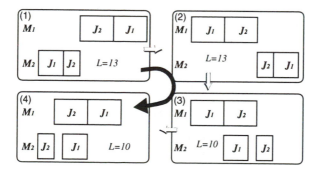

Figure 7.14   Schedule reversal and activation

7.7.5 MSXF-GA for job shop scheduling

The MSXF is applied to the JSSP by using the active CB neighbourhood and the DG distance previously defined. Algorithm 7.7.4 describes the outline of the MSXF-GA routine for the JSSP using the steady state model proposed in [24, 28]. To avoid premature convergence even under a small population condition, an individual whose fitness value is equal to that of someone in the population is not inserted into the population in Step 4.

A mechanism to search in the space of both the left and right active schedules is introduced into the MSXF-GA as follows. First, there are equal numbers of left and right active schedules in the initial population. The schedule q generated from p_0 and p_1 by the MSXF ought to be left (or right) active if p_0 is left (or right) active, and with some probability (0·1 for example) the direction is reversed.

Algorithm 7.7.4 MSXF-GA for the JSSP

- Initialise population: randomly generate a set of left and right active schedules in equal number and apply the local search to each of them.

do 1 Randomly select two schedules p_0, p_1 from the population with some bias depending on their makespan values.

    2 Change the direction (left or right) of p_1 by reversing the job sequences with probability P_r.

    3 Do step 3(a) with probability P_c, or otherwise do step 3(b).

       (a) *If* the DG distance between p_1, p_2 is shorter than some predefined small value, apply MSMF to p_1 and generate q. *Otherwise*, apply MSXF to p_1, p_2 using the active CB neighbourhood $N(p_1)$ and the DG distance and generate a new schedule q.

       (b) Apply algorithm 7.7.1 with acceptance probability given by eqn. 7.2 and the active CB neighbourhood.

    4 If q's makespan is shorter than the worst in the population, and no one in the population has the same makespan as q, replace the worst individual with q.

until some termination condition is satisfied.

- Output the best schedule in the population.

Figure 7.15 shows all of the solutions generated by an application of (a) the MSXF and (b) a stochastic local search computationally equivalent to (a) for comparison. Both (a) and (b) started from the same solution (the same parent p_0), but in (a) transitions were biased toward the other solution p_1. The x axis represents the number of disjunctive arcs whose directions are different from those of p_1 on machines with odd numbers, i.e. the DG distance was restricted to odd machines. Similarly, the y axis representing the DG distance was restricted to even machines.

7.8 Benchmark problems

The two well known benchmark problems with sizes of 10×10 and 20×5 (known as mt10 and mt20) formulated by Muth and Thompson [18] are commonly used as test beds to measure the effectiveness of a certain method. The mt10 problem used to be called a notorious problem, because it remained unsolved for over 20 years; however it is no longer a computational challenge.

Applegate and Cook proposed a set of benchmark problems called the ten tough problems as a more difficult computational challenge than the mt10 problem, by collecting difficult problems from literature, some of which still remain unsolved [3].

7.8.1 Muth and Thompson benchmark

Table 7.2 summarises the makespan performance of the methods described in this Chapter. This Table is partially cited from [6]. The conventional GA has only limited success and is outdated. It would be improved by being combined with the GT algorithm and/or the schedule reversal. The other results, excluding the MSXF-GA results, are somewhat similar to each other, although the SXX-GA is an improvement over the GT-GA in terms of speed and the number of times needed to find optimal solutions for the mt10 problem. The SB-GA produces better results using the very efficient and tailored shifting bottleneck procedure. The MSXF-GA, which combines a GA and local search, obtains the best results.

For the MSXF-GA, the population size = 10, constant temperature c = 10, number of iterations for each MSXF = 1000, $P_r = 0 \cdot 1$ and $P_c = 0 \cdot 5$ are used. The MSXF-GA experiments were performed on a DEC Alpha 600 5/226 which is about four times faster than a Sparc Station 10, and the programs were written in the C language. The MSXF-GA finds the optimal solutions for the mt10 and mt20 problems almost every time in less than five minutes on average.

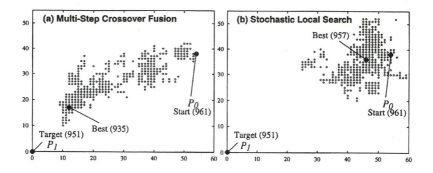

Figure 7.15   Distribution of solutions generated by an application of (a) MSXF and (b) a short-term stochastic local search

Table 7.2   Performance comparison using the MT benchmark problems

1963	Muth-Thompson	test problems	10×10	20×5
1991	Nakano/Yamada	conventional GA	965	1215
1992	Yamada/Nakano	Giffler-Thompson GT-GA	930	1184
	Dorndorf/Pesch	priority rule based P-GA	960	1249
	Dorndorf/Pesch	shifting bottleneck SB-GA	938	1178
1995	Kobayashi/Ono /Yamamura	subsequence exchange crossover SXX-GA	930	1178
1995	Bierwirth	generalised permutation GP-GA	936	1181
1996	Yamada/Nakano	multistep crossover fusion MSXF-GA	930	1165

7.8.2 The ten tough benchmark problems

Table 7.3 shows the makespan performance statistics of the MSXF-GA for the ten difficult benchmark problems proposed in [3]. The parameters used here were the same as those for the MT benchmark except for the population size = 20. The algorithm was terminated when an optimal solution was found or after 40 minutes of CPU time passed on the DEC Alpha 600 5/266. In the table, the column labelled lb shows the known lower bound or known optimal value (for la40) of the makespan, and the columns labelled bst, avg, var and wst show the best, average, variance and worst makespan values obtained, over 30 runs, respectively. The columns labelled n_{opt} and t_{opt} show the number of runs in which the optimal schedules are obtained and their average CPU times in seconds. The problem data and lower bounds are taken from the OR-library [5]. Optimal solutions were found for half of the ten problems, and four of them were found very quickly. The small variances in the solution qualities indicate the stability of the MSXF-GA as an approximation method.

Table 7.3   Results of the ten tough problems

prob	size	lb	bst	avg	var	wst	n_{opt}	t_{opt}
abz7	20 3 15	655	678	692·5	0·94	703	–	–
abz8	20 3 15	638	686	703·1	1·54	724	–	–
abz9	20 3 15	656	697	719·6	1·53	732	–	–
la21	15 3 10	–	*1046	1049·9	0·57	1055	9	687·7
la24	15 3 10	–	*935	938·8	0·34	941	4	864·1
la25	20 3 10	–	*977	979·6	0·40	984	9	765·6
la27	20 3 10	–	*1235	1253·6	1·56	1269	1	2364·75
la29	20 3 10	1130	1166	1181·9	1·31	1195	–	–
la38	15 3 15	–	*1196	1198·4	0·71	1208	21	1051·3
la40	15 3 15	*1222	1224	1227·9	0·43	1233	–	–

Figure 7.16 shows a performance comparison of with and without the MSXF using the la38 problem. A total of 100 runs was done for each under the same conditions used in Table 7.3. The solid line gives the MSXF-GA results and the dotted line gives the equivalent GLS results using a short-term stochastic local search. The y axis shows the CPU time at which each run is terminated and the x axis shows the run numbers which are sorted in ascending order according to the CPU times. The CPU time value = 2400 means that the run was terminated before it found the optimal schedule. The experiments with the MSXF outperformed those without the MSXF both in terms of the CPU time and in the number of successful runs.

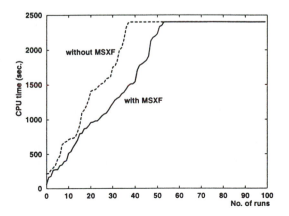

Figure 7.16   Performance comparison using the 1a38 15 × 15 problem

7.9 Other heuristic methods

Local search based meta-heuristics are commonly applied to the JSSP such as simulated annealing (SA) and tabu search (TS). Van Laarhoven *et al.* [16] proposed an SA approach by using the AS neighbourhood described in Section 7.7.3. Matsuo *et al.* proposed a similar SA approach but with more control. Taillard proposed a TS approach that uses the same neighbourhood. Dell'Amico and Trubian extended and improved Taillard's TS method using CB neighbourhood. More recently, Nowicki and Smutnicki [20] proposed a still more powerful TS method. Yamada and Nakano proposed an SA approach combined with the shifting bottleneck and improved the best solutions for the two problems abz9 and la29 of the ten tough problems [33]. Balas and Vazacopoulos proposed the guided local search procedure and combined it with the shifting bottleneck, which at present outperforms most existing methods. For more comprehensive reviews, please refer to [17], [27] and [8].

7.10 Conclusions

The first serious application of GAs to solve the JSSP was proposed by Nakano and Yamada using a bit string representation and conventional genetic operators. Although this approach is simple and straightforward, it is not very powerful. The idea of using the GT algorithm as a basic schedule builder was first proposed by Yamada and Nakano [30] and by Dorndorf and Pesch [12, 22] independently. The approaches by both groups and other active schedule-based GAs are suitable for middle-size problems; however, it seems necessary to combine each with other heuristics such as the shifting bottleneck or local search to solve larger-size problems.

To solve larger-size problems effectively, it was crucial to incorporate local search methods which use domain-specific knowledge. The multistep crossover fusion (MSXF) was proposed by Yamada and Nakano as a unified operator of a local search method and a recombination operator in genetic local search. The MSXF-GA outperforms other GA methods in terms of the MT benchmark and is able to find near-optimal solutions for the ten difficult benchmark problems, including optimal solutions for five of them.

7.11 References

1   Adams, J., Balas, E., and Zawack, D.: 'The shifting bottleneck procedure for job shop scheduling', *Manage. Sci.*, **34**, (3), pp. 391–401, 1988

2   Applegate. D.: *Jobshop benchmark problem set* (Personal Communication, 1992)

3   Applegate, D., and Cook, W.: 'A computational study of the job-shop scheduling problem', *ORSA J. Comput.*, **3**(2), pp. 149–156, 1991

4   Balas, E.: 'Machine sequencing via disjunctive graphs: an implicit enumeration algorithm', *Oper. Res.*, **17**, pp. 941–957, 1969

5   Beasley, J. E.: 'Or-library: distributing test problems by electronic mail', *Oper. Res.*, **41**, pp. 1069–1072, 1990

6   Bierwirth, C.: 'A generalized permutation approach to job shop scheduling with genetic algorithms', *OR Spektrum*, **17**, pp. 87–92, 1995

7   Bierwirth, C., Mattfeld, D., and Kopfer, H.: 'On permutation representations for scheduling problems'. In 4th *PPSN*, 1996

8   Blazewicz, J., Domschke, W., and Pesch, E.: 'The job shop scheduling problem: conventional and new solution techniques', *Oper. Res.*, pp. 1–33, 1996

9   Brucker, P., Jurisch, B., and Sievers, B.: 'A branch & bound algorithm for the job-shop scheduling problem, *Discrete Appl. Math.*, **49**, pp. 107–127, 1994

10 Davidor, Y., Yamada, T., and Nakano, R.: 'The ecological framework II: Improving GA performance at virtually zero cost'. Proceedings of 5th *ICGA*, pp. 171–176, 1993

11 Dell'Amico, M., and Trubian, M.: 'Applying tabu search to the job-shop scheduling problem', *Ann. Oper. Res.*, **41**, pp. 231–252, 1993

12 Dorndorf, U., and Pesch, E.: 'Evolution based learning in a job shop scheduling environment', *Comput. Oper. Res.*, **22**, pp. 25–40, 1995

13 Giffler, B., and Thompson, G.L.: 'Algorithms for solving production scheduling problems', *Oper. Res.*, **8**, pp. 487–503, 1960

14 Yamamaru, M., T., Ono, and Kobayashi, S.: 'Character-preserving genetic algorithms for traveling salesman problem', (in Japanese), *J. Jpn. Soc. Artif. Intell.*, **7**, pp. 1049–1059, 1992

15 Kobayashi, S., Ono, I., and Yamamura, M.: 'An efficient genetic algorithm for job shop scheduling problems'. Proceedings of 6th *ICGA*, pp. 506–511, 1995

16 van Laarhoven, P.J.M., Aarts, E.H.L., and Lenstra, J.K.: 'Job shop scheduling by simulated annealing', *Oper. Res.*, **40**,(1), pp. 113–125, 1992

17 Mattfeld, D. C.: *Evolutionary search and the job shop; investigations on genetic algorithms for production scheduling* (Physica Verlag, Heidelberg, Germany, 1996)

18 Muth, J.F., and Thompson, G.L.: *Industrial scheduling* (Prentice-Hall, Englewood Cliffs, NJ, 1963)

19 Nakano, R., and Yamada, T.: 'Conventional genetic algorithm for job shop problems' Proceedings of 4th *ICGA*, pp. 474–479, 1991

20 Nowicki, E., and Smutnicki, C.: 'A fast taboo search algorithm for the job shop problem', *Manage. Sci.*, 1995

21 Panwalkar, S. S., and Iskander, W.: 'A survey of scheduling rules', *Oper. Res.*, **25**, (1), pp. 45–61, 1977

22 Pesch, E.: *Learning in automated manufacturing: a local search approach* (Physica-Verlag, Heidelberg, Germany, 1994)

23 Reeves, C., R.: 'Genetic algorithms and neighbourhood search', in *Evolutionary computing, AISB workshop (Leeds, UK)*, pp. 115–130, 1994

24 Syswerda, G.: 'Uniform crossover in genetic algorithms'. Proceedings of 3rd *ICGA*, pp. 2–9, 1989

25 Taillard, E.D.: 'Parallel taboo search techniques for the job-shop scheduling problem', *ORSA J. Comput.*, **6**,(2), pp. 108–117, 1994

26 Ulder, N.L.J., Pesch, E., van Laarhoven, P.J.M., Bandelt, H. J., and Aarts, E.H.L.: 'Genetic local search algorithm for the traveling salesman problem', proceedings of 1st *PPSN*, pp. 109–116, 1994

27 Vaessens, R.J.M.: 'Generalized job shop scheduling: complexity and local search'. Dissertation, University of Technology Eindhoven, 1995

28 Whitley, D.: 'The genitor algorithm and selection pressure: why rank-based allocation of reproductive trials is best'. Proceedings of 3rd *ICGA*, pp. 116–121, 1989

29 Yagiura, M., Nagamochi, H., and Ibaraki, T.: 'Two comments on the subtour exchange crossover operator', *J. Jpn. Soc. Artif. Intell.*, **10**, pp. 464–467, 1995

30 Yamada, T., and Nakano, R.: 'A genetic algorithm applicable to large-scale job-shop problems', proceedings of 2nd *PPSN*, pp. 281–290, 1992

31 Yamada, T., and Nakano, R.: 'A genetic algorithm with multi-step crossover for job-shop scheduling problems'. Proceedings of first IEE/IEEE international conference on *Genetic algorithms in engineering systems: innovations and applications, GALESIA '95*, pp. 146–151, 1995

32 Yamada, T., and Nakano, R.: 'A fusion of crossover and local search'. Proceedings of IEEE international conference on *Industrial technology (ICIT '96)*, 1996

33 Yamada, T., and Nakano, R.: *Job-shop scheduling by simulated annealing combined with deterministic local search* (Kluwer academic publishers, MA, USA, 1996)

34 Yamada, T., and Nakano, R.: 'Scheduling by genetic local search with multi-step crossover. Proceedings of 4th *PPSN*, 1996

35 Yamada, T., Rosen, B.E., and Nakano, R.: 'A simulated annealing approach to job shop scheduling using critical block transition operators'. Proc. IEEE int. conf. on *Neural Networks, Orlando, Florida*, pp. 4687–4692, 1994

Chapter 8
Evolutionary algorithms for robotic systems: principles and implementations

A. M. S. Zalzala, M. C. Ang, M. Chen, A. S. Rana and Q. Wang

This Chapter addresses the principles of the use of evolutionary algorithms in the motion planning of robotic systems. In addition, the implementation of these principles is then reported for single manipulators, multiple arms and mobile arms.

A general form of the robot control loop is shown in Figure 8.1. The required job is first divided by the task planner producing a number of consecutive tasks, followed by the motion planner which gives a time history of positions, velocities and accelerations sufficient and necessary to realise each task. Once the desired motion elements are available, they are used to produce the commands for the individual joint loops via the control module, which may or may not include the inverse model of the system. The motion is realised by applying the control commands to the robot system and a feedback module provides the actual motion elements to cater for any uncertainties and/or changes in the system parameters and/or environment set up. The algorithms reported in this Chapter are mainly concerned with the motion planning block of Figure 8.1.

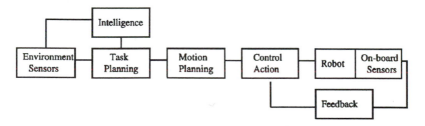

Figure 8.1   Overview of a robot integrated system

Although the basic ideas behind a genetic algorithm are well defined, particularly for binary-based populations, the deployment of an evolutionary scheme to provide for a solution to particular problems is far from being straightforward. In robot-related applications, the choice of operators and objective formulation, let alone the initial coding of the problem, are of vital importance. However, the design of these evolutionary components may vary amongst problems (e.g. planning or control) and systems (e.g. multi-dimensional articulated chains or 2D mobile vehicles).

The following Sections report on work carried out within the Robotics Research Group at the University of Sheffield aiming to investigate the design of evolutionary structures for the optimisation of robot operations. Although the choice of an evolutionary scheme is considered as a problem-related approach based on both system and environment requirements, the basis of a successful evolutionary algorithm is identified as the appropriate choice of genetic coding, genetic parameters and optimised objective function. The basic assumptions on genetic structures remain valid [8], but extra (or modified) parameters, more efficient chromosome representations and problem-tailored cost functions prove fundamental in the provision for accurate and efficient implementation.

8.1 Optimal motion of industrial robot arms

Trajectory planning of manipulators requires the provision of a time history of motion for the arm to accomplish a specific task. However, there are infinite trajectories, in the joint space, for a robotic manipulator to move from one position to another, and a decision should be made on which trajectory to use according to some specified criteria. In addition to other criteria, the motion may be optimised considering the travel time, energy consumption and/or environmental constraints. Nonetheless, the need to combine more than one criteria in the optimisation process may prove difficult due to the often conflicting natures of the considered criteria.

Optimum trajectory planning was attempted by many researchers, with some of the early works on minimum-time motion using either a simplified dynamic model (e.g. [1]) or no dynamics at all (e.g. [2]). This continued to be the trend in most of the literature, with constraints on motion consisting of purely kinematic bounds on positions and their derivatives. However, a joint-space tessellation and a graph search scheme was presented [3], planning for optimal-time motion via an exhaustive search method. The full dynamic models of the manipulator and actuator torque limits were both taken into consideration in arriving at the time-optimal trajectory. However, only a two-joint arm was considered and the authors reported a vast increase in the search time when the tessellated grid increases in size. This approach in [3] appears to be the ultimate solution for this type of planning problem, but seems to have been rendered inefficient by the excessive computations involved.

In robotics, GAs have mainly been used in path planning and decision making on collision avoidance. The GAs' randomised, although structured, exchange mechanism exploits historical information to speculate on new search points with expected improved performance. The direction of the search is influenced only by the objective function associated with the individuals' fitness levels. GAs search for optimum solutions globally, thus avoiding getting trapped in a local minimum considered as a common handicap in conventional methods [4]. Other benefits include GAs' feasibility to be paralleled on a distributed multiprocessor system [5], thus providing for one main requirement in a real-time implementation.

8.1.1 Formulation of the problem

Manipulator trajectories consist of finite sequences of positions (joint angles) and it is suitable to code these into a string of the format:

$$\begin{bmatrix} \theta_{11} & , \theta_{12} & , \bullet & \bullet & \bullet & , \theta_{1n} \\ \theta_{21} & , \theta_{22} & , \bullet & \bullet & \bullet & , \theta_{2n} \\ & & \bullet & \bullet & \bullet & \bullet & \bullet \\ \theta_{m1} & , \theta_{m2} & , \bullet & \bullet & \bullet & , \theta_{mn} \end{bmatrix}$$
$$\begin{bmatrix} swpos \end{bmatrix}$$

where θ_{ij} is the jth intermediate position node of the ith link, and n is the number of intermediate position nodes and is 16 in this simulation. The value *swpos* is the switch position (its decimal value varies from 1 to $n-1$) information chromosome, which is used for the heuristic search. The objective is to minimise the travel time. The end velocities should ideally be zeros if the arm is required to rest at the end of the motion. Thus, these velocity constraints are incorporated as penalties, resulting in the following objective function:

$$J = \sum_{j=1}^{n} h_j + \lambda_v \sum_{i=1}^{6} |v_n^i|$$

where h_j is the time interval for the jth segment of motion and is calculated by the dynamic scaling scheme and λ_v is a weighting factor (equal to 0·1 in this application). The fitness of a chromosome is denoted by:

$$fitness = 2{\cdot}0 - \frac{J}{\max J}$$

where J is the objective function value and is the maximum objective value in the same generation of populations. This definition ranges the fitness from 1·0 to 2·0, and the chromosome is then made to perform reproduction, crossover and mutation.

Rather than minimising time, some applications require the minimisation of torque values applied to the arm's motors. In this case, a new weighting factor, λ_v, is introduced with a value of 0·06. Thus, a combined objective function was designed as:

$$OBJ = \sum_{j=1}^{n} h_j + \lambda_\tau \sum_{j=1}^{n} \tau_j + \lambda_v \sum_{i=1}^{6} \left| v_n^i \right|$$

During reproduction, the number of occurrences of the same trajectories selected for crossover is limited, which encourages higher interaction among different trajectories. To prevent any path dominating the population and leading to premature convergence, only a specific number of copies of the same trajectory are allowed to remain in the population after reproduction, and extra copies are replaced by new trajectories.

Single point crossover was adopted as an initial study. After choosing a crossing site in one parent string, crossover is performed only if the crossover site of the second parent is within certain proximity of the circle centred at the first crossing site. In all, ten crossover sites are randomly generated and checked to meet this restriction; if none does, then crossover fails.

In the following example, the crossover rate is chosen as 0·8, i.e. 80 percent of the pairs are going to do crossover. Let two parents have the following formats:

$$
\begin{bmatrix}
q_{11}, q_{12}, \ldots, q_{16}, \ldots, q_{1n} \\
q_{21}, q_{22}, \ldots, q_{26}, \ldots, q_{2n} \\
\ldots \quad\quad .. \quad\quad \ldots \\
q_{m1}, q_{m2}, \ldots, q_{m6}, \ldots, q_{mn}
\end{bmatrix}
\qquad
\begin{bmatrix}
\theta_{11}, \theta_{12}, \ldots, \theta_{16}, \ldots, \theta_{1n} \\
\theta_{21}, \theta_{22}, \ldots, \theta_{26}, \ldots, \theta_{2n} \\
\ldots \quad\quad .. \quad\quad \ldots \\
\theta_{m1}, \theta_{m2}, \ldots, \theta_{m6}, \ldots, \theta_{mn}
\end{bmatrix}
$$
$$
[1101] \qquad\qquad\qquad [1011]
$$

and suppose the crossover position for the trajectory is randomly chosen as 6 and the crossover position for the switch position *swpos* is 2 (can be between 1 and 4). As a result, the two offspring have the following forms:

$$
\begin{bmatrix}
\theta_{11}, \theta_{12}, \ldots, q_{16}, \ldots, q_{1n} \\
\theta_{21}, \theta_{22}, \ldots, q_{26}, \ldots, q_{2n} \\
\ldots \quad\quad .. \quad\quad \ldots \\
\theta_{m1}, \theta_{m2}, \ldots, q_{m6}, \ldots, q_{mn}
\end{bmatrix}
\qquad
\begin{bmatrix}
q_{11}, q_{12}, \ldots, \theta_{16}, \ldots, \theta_{1n} \\
q_{21}, q_{22}, \ldots, \theta_{26}, \ldots, \theta_{2n} \\
\ldots \quad\quad .. \quad\quad \ldots \\
q_{m1}, q_{m2}, \ldots, \theta_{m6}, \ldots, \theta_{mn}
\end{bmatrix}
$$
$$
[1111] \qquad\qquad\qquad [1001]
$$

Mutation is practised in such a way as to slightly alter the position of one via point on the trajectory, as shown in Figure 8.2. Thus, the current position, q_j, is altered by offset ds (set to 0·0125 rad) according to the mutation rate, with the direction of variation randomly chosen. The mutation is repeated for every element of the string matrix.

Figure 8.2   The mutation scheme

8.1.2 Simulation of case studies

8.1.2.1 A two DOF arm

The manipulator considered here is a classical two joint planar arm. In this simulation, the optimisation of time with end point velocity penalty is considered, and the effect of gravitation is not taken into account. The required planning is for the motion of joint one moving from 0 to 1 and joint two moving from –2 to –1 (radians). The maximum generation size was set to 600, and the best objective found was 0·4046 seconds at iteration 462. The best switch position, *swpos*, was near the middle of the time axis of the grid. Figure 8.3 shows the history of the objective value against the number of generations. Figures 8.4–8.6 show the near time-optimal trajectories searched by the GA in position, velocity and acceleration profiles, respectively, where the solid lines are for joint one and the dashed lines for joint two. Figure 8.7 shows all joints within their limit boundaries, with joint one only enjoying bang-bang motion.

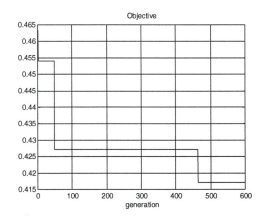

Figure 8.3   Objective history

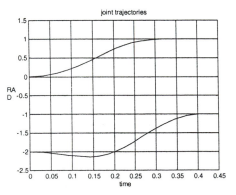

Figure 8.4   Minimum-time path

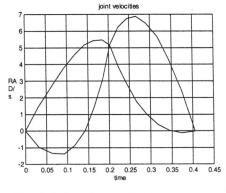

Figure 8.5   Joint velocities

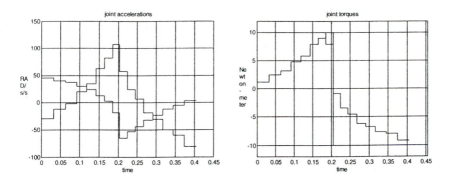

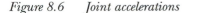

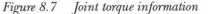

Figure 8.6   *Joint accelerations*   Figure 8.7   *Joint torque information*

The performance of the proposed genetic planner is compared against another method reported in the literature [6]. For all four cases suggested in [6], the genetic planner outperformed the other, as shown in Table 8.1.

Table 8.1 *Comparing travel time values with Dissanayake (1991)*

	Initial positions (rad)		Final positions (rad)		Travel time results (sec)	
	Θ_{10}	Θ_{20}	Θ_{1f}	Θ_{2f}	Dissanayake [6]	this chapter
Motion case 1	0·00	−2·00	1·00	−1·00	0·6711	0·4046
Motion case 2	1·00	−1·00	0·00	−2·00	0·6732	0·4123
Motion case 3	1·32	−2·64	2·80	−2·37	0·6404	0·3970
Motion case 4	2·80	−2·37	1·32	−2·64	0·6185	0·3927

8.1.2.2 A six DOF arm

In this Section, simulation results are reported for the PUMA 560 arm considering the different objective functions given in Section 8.1.1. The start and end positions used in all seven case studies for all six joints are given in Table 8.2.

Table 8.2   *The motion start and end points (radians)*

	Joint 1	Joint 2	Joint 3	Joint 4	Joint 5	Joint 6
Start	−0·3	0·4	−0·18	0·0	−0·05	0·05
End	0·51	−0·42	0·58	0·87	0·64	0·84

The GA parameters are chosen as follows:

- The maximum number of similar members in the new population, *criteria* = 4.
- The absolute number of grid variations allowed during an intermediate position mutation, *variation* = 2.
- The maximum number of similar members allowed to crossover during a reproduction, *samemax* = 6.
- The crossover rate, *xor* = 0·80.
- The mutation rate for varying intermediate point co-ordinates, *pmutr* = 0.10.
- The maximum generation *na* = 300.

The first case study considers time optimisation with velocity constraints, and the best objective value was 0·431 seconds at iteration 70. To provide for a study of the effect of using combined optimisation criteria in robot motion planning, other case studies are included along with the above, as shown in Table 8.3. One important parameter in any algorithm using grid search is the actual size of the grid representing the searched space, as the complexity of the search increases exponentially with the number of points in a chosen grid. Thus, it is always sensible to have a certain trade off between search resolution and computation time (note case 7 in Table 8.3). As expected, all cases where the optimisation is constrained by a near-zero end-point velocity exhibit a higher motion time. The presence of such constraint is important, however, if motion is to be planned via successive segments, as is the case for a point-look-ahead motion planner. The motion profiles for all six joints of the PUMA are shown in Figure 8.8 for case 5 of Table 8.3.

Table 8.3 Simulation results of different case studies for the PUMA

Case no. λv	Grid size	Optimisation criteria			Parameters $\lambda_T + \mp +$		Motion	No. of
		time	torque	velocity				
		timegenerations						
				constraints			(s)	
1	16 3 16	yes	no	yes	—	0·1	0·431	70
2	16 3 16	yes	no	no	—	—	0·475	255
3	16 3 16	no	yes	yes	0.06	0·1	0·502	248
4	16 3 16	no	yes	no	0.06	—	0·307	220
5	16 3 16	yes	yes	yes	0.06	0·1	0·873	280
6	16 3 16	yes	yes	no	0.06	—	0·367	285
7	25 3 25	yes	no	yes	—	0·1	0·421	40 000

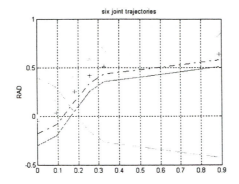

Figure 8.8   Near minimum time PUMA motion (position profile)

8.1.3 Parallel genetic algorithms

The main motivation for exploring parallel GAs is to improve speed and efficiency by employing a parallel computer. Migration GAs divide the population into a number of subpopulations, each of which is treated as a separate breeding unit under the control of a conventional GA. To encourage the proliferation of good genetic material throughout the whole population, good individuals migrate between the subpopulations from time to time. In natural evolution, species tend to reproduce within subgroups. Although it is expected that individuals from within the same group will reproduce together, occasional migration of individuals occurs between subpopulations such that individuals from one population are introduced into another subpopulation. Using a set of three processors to house three separate subpopulations, a duplicate of the best chromosome of each servant group is immigrated to the root group, and the less important chromosome on the root processor is replaced. The semi-isolation of subpopulations and limited communication between them also encourages a high degree of fault tolerance along with efficient utilisation of processors. Table 8.4 indicates the superiority of a three-group parallel GA as compared to a sequential GA.

Table 8.4   Performance comparison

Approach	Generations no.
Sequential GA	2000
Parallel GA on 3 transputers	70

8.2 A comparative study of the optimisation of cubic polynomial robot motion

This Section is concerned with the motion optimisation of a SCARA robot subject to kinematics constraints, where two methods are used, namely: genetic-based algorithms and the flexible polyhedron search. Both methods use cubic spline functions to generate motion profiles. In formulating the GAs, tailor-fit operators and procedures were used to seek an application-dependent structure. In addition, an initial evaluation is reported in the form of comparisons between a Pareto-based and weighted-sum (parametric) approaches to multicriteria optimisation. Case study results are summarised for the RTX robot with six joints.

8.2.1 Background

Robotic motion planning is an optimisation problem, in which a robot path or trajectory has to be planned based on different optimisation criteria. Due to the competition between various optimisation criteria, a multicriteria motion planning problem typically displays many local minima, and conventional optimisation methods often fail to tackle it. Genetic algorithms search for a solution through a population and may therefore avoid being trapped in local minima.

The application of GAs in multiobjective optimisation is a major interest in GA research. Current multiobjective GA approaches include the classical aggregation of the different objectives into a single function, population-based non-Pareto approaches and the most recent work on the ranking schemes based on the definition of Pareto optimality [7]. Pareto-based ranking was first proposed by Goldberg [8], as a means of assigning equal probability of reproduction to all Pareto optimum[1] individuals. However, this is not the only technique required for the Pareto-based multiobjective GAs. There are additional niche formation methods which must be included to prevent the population converging to one peak, a phenomenon know as genetic drift, as described below.

The flexible polyhedron search is one of the unconstrained non-linear optimisation techniques [9]. It is used for the minimisation of a function of n variables and depends on the comparison of function values at the $n+1$ vertices, followed by the replacement of the vertex with the highest value by another point. This method is widely used by researchers since it does not assume the function to be differentiable and continuous over the range of interest, and has the additional

[1] Also called efficient point, nondominated or admission solution.

advantage that the user can be sure of the design variables taking only positive values. This method has been adopted [2] to optimise the best (minimum) combination of time intervals subject to constraints on joint velocities, accelerations and jerks for a PUMA type robot with six joints.

8.2.2 Motion based on cubic splines

The objective here is to construct joint trajectories which fit a number of joint displacements at a sequence of time instants by using cubic polynomial functions. Further, the motion has to comply with the maximum bounds on position and its derivatives. Thus, the multicriteria objective to optimise is:

$$\lambda = \max\left(1, \lambda_1, \sqrt[2]{\lambda_2}, \sqrt[3]{\lambda_3}\right)$$

The derivation and use of the above equation is explained in [2] and included briefly in the Appendix for completion.

8.2.3 The genetic formulations

The formulation of the multiobjective GAs is described in some details in the following Sections. The general procedure is depicted in Figure 8.9.

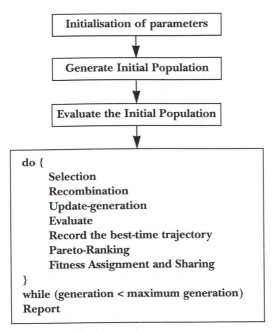

Figure 8.9   Outline of the genetic procedure

8.2.4 The objective functions

8.2.4.1 Pareto-based GA

The objective vector is set as:

$$\text{minimise } \{ \sum_{i=1}^{n-1} h_i,\ 1-\lambda_1,\ 1-\lambda_2 \} \text{ subject to constraint: } \lambda_3 \leq 1$$

where h_i is the ith time interval, $1-\lambda_1$ is the criticality to the velocity constraints, $1-\lambda_2$ is the criticality to acceleration constraints. The criticality is a measure of how close a trajectory is to the joints' velocity and acceleration limits.

8.2.4.2 Weighted-sum GA

Two types of weighting methods are considered in this case study, namely a normalised objective function and a non-normalised function.
(*a*) The formulation of objective function in normalised form can be written as:

$$\text{minimise } \{ w_1 f_1 + w_2 f_2 + w_3 f_3 \}$$

subject to jerk constraints, i.e. $\lambda_3 \leq 1$, where $f_2 = 1-\lambda_1$ and $f_3 = 1-\lambda_2$. The weight coefficients w_i operate on the ith objective function and can be interpreted as the relative worth of that objective when compared to the other objectives. The weights are normalised so that the total is equal to 1, and the two objective values for f_2 and f_3 are in the range [0,1].

The objective function f_1 is calculated as a relative value to the range of motion time [*MIN, MAX*]. Assume the upper bound of travel time, *MAX*, is the maximum trajectory motion time obtained from the initial population, and the lower bound of the travel time, *MIN*, is the maximum time among the six joints when they travel at their velocity limits between start and end points, i.e. θ_{j1} and $\theta_{j,n}$. Thus:

$$MIN = \max_{j} \left(\frac{\theta_{jn} - \theta_{j1}}{VC_j} \right)$$

where $j = 1,..., N$ represent the six joints and VC_j is the velocity constraint imposed on joint j. Note that the lower bound of the motion time calculation has not taken into account the zero velocity and acceleration values imposed at the start and end positions. Hence, the function f_1 is set as:

$$f_1 = \frac{\sum\limits_{i=1}^{n-1} h_i - MIN}{MAX - MIN}$$

(*b*) Some researchers have not adopted the concept of normalising the objectives when using weighting methods in their studies [3,4], and the objective functions are arbitrarily weighted with different values. To compare with the normalised formulation, the non-normalised objective function is formulated as:

$$\text{minimise } \{ \omega_1 \sum\limits_{i=1}^{n-1} h_i + \omega_2 f_2 + \omega_3 f_3 \}$$

subject to jerk constraints, where ω_i is the weight of objective f_i.

8.2.5 Parameter initialisation

The definitions of the used genetic parameters are listed in Table 8.5, and these are referred to in the following Sections.

Table 8.5   The GA parameters

Parameter	Description
maxgen	maximum generation
maxinterval/(n-1)	maximum time interval
maxjoint	maximum joint number
maxknot/n	maximum knot number
maxpop	maximum population size
SEED	constant integer to initialise the random number generator
sp	selective pressure
pcross	crossover probability
pinject	injection probability
pmutate	BGA (breeder genetic algorithm) mutation probability
pmutate_time	time mutation probability
reserve_num	number of best trajectory to be kept, it is set to 1, (when performing BGA with truncation scheme, it is also the number of selected parents)
cross_over_dis_meter	maximum distance to perform crossover for joint 1 (Zed)
cross_over_dis_radian	maximum distance to perform crossover for joints 2–6
v1	velocity at the start position
a1	acceleration at the start position
vn	velocity at the end position
an	acceleration at the end position
sizeC	the number of constraints
inc	time step for generating motion profiles

8.2.6 Evaluating the population

The population performance is evaluated with each trajectory in the population receiving a fitness value prior to the selection process.

8.2.6.1 Ranking

All trajectories are ranked based on their total travelling time and how critical they are to the joints' velocity and acceleration limits, based upon Pareto ranking. Using the Pareto optimality definition, a point x is said to be nondominated if not dominated by any other point. Thus, an individual can be ranked by counting the number of other individuals dominating it [7]. Hence, the nondominated trajectories that are the best performers will be assigned with highest rank, i.e. zero. When all individuals in the population are ranked, the fitness values will be assigned to them according to their rank. This can be done by interpolating some linear or exponential function from the best individual (rank = 0) to the worst individual (rank < N). Following that, same rank individuals will receive the same fitness values by averaging the total values assigned to them.

   Therefore, fitness values are assigned according to an individual's rank in the population, thus ensuring that the population will strive for all the three objectives: smallest travel time and achieving the two limits allowable.

8.2.6.2 Fitness assignment

The fitness of trajectories is obtained on the basis of their relative fitness in the population rather than their raw performances. The trajectories are first sorted into a descending order based on their Pareto rank. Then by interpolating between the best ranked individual and the worst ranked individual, each trajectory fitness value can be calculated as follows:

$$F(x_i) = 2 - sp + 2(sp - 1)\frac{x_i - 1}{Maxpop - 1}$$

where x_i is the position in the ordered population of trajectory i [10]. This rank-based fitness assignment provides only a small bias towards the most fit trajectories so that no trajectory will generate an excessive number of offspring and thus prevent premature convergence. sp is the selective pressure and defines the maximum number of offspring that the best trajectory can reproduce.

8.2.6.3 Sharing scheme

Fitness sharing uses a sharing parameter to control the extent of sharing, or in other words, it is a measure of the maximum distance between individuals which could form niches [7]. Trajectory fitness will be increased or decreased depending on similarity with other trajectories. After calculating the share count, the new total fitness values in the population will be altered and the value is usually different from the total fitness before share count.

8.2.7 Selection scheme

Selection scheme is a process for determining the number of trials for which a particular individual is chosen for reproduction. The selection technique adopted in this project is based on stochastic universal sampling (SUS) [11]. This method uses a single spin and N equally spaced pointers, where N is the population size. The actual selection process begins with the generation of a random number, say p, from the range $[0, sum/N]$, of which sum is the total fitness values of the population. The N trajectories are then chosen by generating the N pointers spaced by sum/N. Hence, N trajectories, whose fitnesses span the positions of the pointers, will be selected. The selected trajectories are then shuffled randomly before recombination. Note, however, that the number to be selected for this particular problem is not equal to N and a modification has been introduced.

8.2.8 Shuffling

After the selection stage, the selected individuals' indexes are sorted because of the previous fitness assignment procedure. It is therefore necessary to perform a shuffling procedure before the recombination process. This can be done by using a set of randomly generated numbers. By using the sorting procedure to sort the random numbers into an ascending order, the initially sorted indexes will be randomised and the selected individuals will be rearranged according to the randomised indexes.

8.2.9 Recombination mechanisms

The selected trajectories will be paired up for crossover or recombination subject to their mating distance and crossover probability. Self mating is not allowed in the program. Three different genetic operators are implemented from the Breeder GA [12].

Considering two parents $\mathbf{x} = (x_1, ...,x_n)$ and $\mathbf{y} = (y_1, ...,y_n)$, then the offspring $\mathbf{z} = (z_1, ...,z_n)$ may be composed in the following ways:

(a) Discrete recombination: $z_i = \{x_i\}$ or $\{y_i\}$, where x_i or y_i are chosen with probability 0.5.
(b) Extended intermediate recombination: $z_i = x_i + \alpha_i (y_i - x_i)$, $i = 1,...,n$, where α_i is chosen randomly in the interval $[-0.25, 1.25]$.
(c) Extended line recombination: $z_i = x_i + \alpha_i (y_i - x_i)$, $i = 1,...,n$, where α is chosen randomly in the interval $[-0.25, 1.25]$.

In addition, a path redistribution/relaxation operator is used [13]. A robot trajectory consists of joint angles which may produce a large position jump in the offspring strings after conventional crossover. Therefore, this operator is proposed where cubic splines are fitted to the offspring's via points with each time interval set to unity. The path length is then computed as the Euclidean distance between the start and end via points along the splines as:

$$\sum_{t=0}^{t=n-\Delta t}\sqrt{\left(\sum_{j=1}^{m}\left(\theta_{j,t+\Delta t} - \theta_{j,t}\right)^2\right)}$$

where $j = 1, ..., m$ and Δt is chosen to be a small number, n is the number of via points and m is the number of robot joints. Joint knots are redistributed evenly over these splines at equal intervals. The paths are then relaxed by moving each via point by a small step towards the point which will bisect the line between its neighbouring points as: where $i = 1, ..., n$ and δ is a positive random number less than one.

8.2.10 Modified feasible solution converter

$$\theta_{ji} = \theta_{ji} + \delta \left(\frac{\theta_{j,i-1} + \theta_{j,i+1}}{2} - \theta_{ji}\right)$$

When a travel time is produced following the process of population generation or recombination, the time value is evaluated using the feasible solution procedure, as described in Section 8.2.3. Then, the total time intervals are scaled such that the motion time of an individual trajectory is optimal and does not violate the kinematics constraints. Thus, the optimisation is re-stated as:

$$\lambda = \max\left(\lambda_1, \sqrt[2]{\lambda_2}, \sqrt[3]{\lambda_3}\right)$$

If $\lambda > 1$ then h_i should be increased to λh_i to satisfy the limits of the velocities, accelerations and jerks. Also, if $0 < \lambda < 1$ then h_i can be contracted to λh_i to obtain the time-optimal joint paths. The corresponding velocity, acceleration and jerk values are scaled by the factors of $\frac{1}{\lambda}$, $\frac{1}{\lambda^2}$, $\frac{1}{\lambda^3}$, respectively.

8.2.11 Time intervals mutation

The offspring's initial time is given from the parent and it will be passed through the modified version of feasible solution converter $n-181$
1 times. Each pass, a time interval will be selected randomly and increased or decreased depending on the criticality of that interval. If the interval is critical, i.e. very close to the limit, the time will be increased (to slow down), otherwise the time will be decreased (to speed up). The decrement and increment values are selected randomly between ranges of $[0 \cdot 75, 1]$ and $[1, 1 \cdot 25]$, respectively. Only the smallest time intervals will be used as the offspring travelling time. If n-1 trials are not successful, the offspring will accept the parental trajectory travel time, scaled appropriately with the factor obtained via the converter.

8.2.12 Simulation results

The algorithms have been tested for the RTX robot with six-joint motion planning in the configuration space. The case study is listed in Table 8.6, with the model parameters and constraints listed in Tables 8.7 and 8.8.

Table 8.6   Initial and final configurations

	Column (m)	Shoulder (rad)	Elbow (rad)	Yaw (rad)	Pitch (rad)	Roll (rad)
Initial configuration	0·4	$-\pi/6$	$-\pi/3$	$-\pi/2$	0	$-\pi/4$
Final configuration	0·8	$\pi/6$	$\pi/3$	$\pi/2$	$-\pi/6$	$\pi/4$

Table 8.7   Link parameters

Joint i	θ_i	α_i	a_i	d_i	Lower bound	Upper bound
1	0°	0°	0	0	-61 mm	+881 mm
2	0°	0°	a_2	$-d_2$	-90°	+90°
3	0°	0°	a_3	$-d_3$	-180°	+151°
4	0°	0°	0	0	-110°	+110°
5	0°	90°	0	0	-8°	+94°
6	90°	90°	0	d_6	-132°	+181°
gripper	-	-	-	-	0°	+90°

Table 8.8   Velocity, acceleration and jerk constraints

	Zed²	Shoulder	Elbow	Yaw	Pitch	Roll
Velocity	0.1116	0.1654	1.2092	1.9715	1.3780	1.2412
Acceleration	1.7755	6.2018	14.081	31.055	28.063	26.180
Jerk	297.59	894.67	3718.9	3377.6	3933.1	4172.7

Note: The zed velocity, acceleration and jerk are in m/s, m/s² and m/s³, respectively. The other joint angle velocities, accelerations and jerks are in rad/s, rad/s² and rad/s³, respectively.

8.2.12.1 Case 1: Pareto-based GA

(a) Breeder genetic algorithm (BGA) operators
The three BGA recombination operators, namely, discrete, extended intermediate and extended line recombination operators, are tested to determine the best among them. Crossover probability is set to (*pcross* = 0·9). The performance for BGA genetic operators is shown in Table 8.9. The results are obtained by using a population size of 200, with 100 as the maximum number of generations.

Table 8.9   Results from BGA genetic operators

	Minimum time(s)
Discrete recombination	4·3381
Intermediate recombination	4·1497
Line recombination	4·0854

(b) Path redistribution-relaxation operator
The path redistribution-relaxation operator is used with a different population size, and 0.9 crossover probability. The injection rate adopted [13] is 10 % of the population size while in this simulation, the injection rate is 2.5 %. The results for 100 generations with different population size are shown in Table 8.10.

Table 8.10   Results for path redistribution-relaxation operator

Population size	Minimum time(s)
100	3·9530
200	3·9335
300	3·9049

² zed is also known as column.

(c) Comparisons

From the above case studies, one can conclude that the extended line recombination operator is the best among the three BGA genetic operators and that the path redistribution-relaxation operator performs better than the BGA recombination operators. A further simulation is carried out for 300 population size and 500 generations with crossover probability of 0·9 and injection of eight new trajectories in each generation. The results are shown in Table 8.11, indicating improved optimisation with larger population and generation for both BGA and redistribution-relaxation operators. However the results also show that the redistribution-relaxation operator still produces the best motion time (smallest) with bigger population and generation.

Table 8.11   Path redistribution-relaxation operator versus BGA recombination operators

Path redistribution-relaxation	Line recombination	Discrete recombination	Intermediate recombination
3.8743 sec	3.8894 sec	4.0144 sec	4.0404 sec

(d) Truncation selection

To complete the study of the BGA operators, truncation selection has been tested with the consideration of sharing but without mating restriction. Hence, T % best rank individuals are selected and mated randomly in each generation to produce the new population. The smallest time trajectory will remain in the population. The results shown in Table 8.12 are obtained with a 200 population size and 100 maximum generation. T % is chosen in the range of 10–50 % [12].

Table 8.12   Performance of the BGA operators without mating restriction

Truncation threshold percentage(T %)	Discrete recombination (s)	Intermediate recombination (s)	Line recombination (s)
10	4·0404	4·0884	4·0373
20	4·4436	4·1031	4·0817
30	4·2443	4·2164	4·0648
40	4·2344	4·2160	4·1575
50	4·7487	4·1698	4·1319

The results above show that, by using truncation selection, the motion time is better than without it if the threshold percentage is small for recombination (note Table 8.9). However, the time obtained by using the redistribution-relaxation operator is still better than that by truncation selection. Table 8.12 also shows again that extended line recombination is indeed the best BGA operator for the minimum time motion planning problem for a nonmating environment.

8.2.12.2 Case 2: Pareto-GA versus flexible polyhedron search

The optimal paths reported in Table 8.11 are fed into the FPS program, and the outcome is given in Table 8.13. Although the GA yielded better optimisation of time, the procedure did require longer computation time than with FPS.

Table 8.13   Pareto-based GA versus flexible polyhedron search

Operator responsible for the optimal path	Minimum time(s)	Results from flexible polyhedron search method(s)
Path redistribution-relaxation	3·8743	4·4870
Line recombination	3·8894	4·1253
Discrete recombination	4·0144	4·1325
Intermediate recombination	4·0404	4·1141

8.2.12.3 Case 3: weighted-sum GA

(*a*) *Normalised formulation*
Several combinations of weights have been tried and the results for a 200 population size for 100 maximum generation are listed in Table 8.14. Only the path redistribution-relaxation operator has been applied in the tests for the weighted-sum GA.

Table 8.14   Results for normalised formulation

Weights(w_1, w_2, w_3)	Minimum time(s)
(0·1,0·1,0·8)	3·9804
(0·2,0·2,0·6)	4·0106
(0·2,0·6,0·2)	4·0044
(0·6,0·2,0·2)	4·0124
(0·05,0·15,0·8)	4·0121
(0·1,0·0,0·9)	3·9531
(0·1,0·8,0·1)	4·0109
(0·8,0·1,0·1)	3·9673
(0·3,0·3,0·4)	3·9570
(0.3,0·4,0·3)	3·9628
(0·4,0·3,0·3)	3·9875
(0·5,0·2,0·3)	3·9807

From the results shown in Table 8.14, when compared to the redistribution-relaxation operator of Pareto-based GA under the same population and generation size, the optimum motion time produced by Pareto-based GA is slightly more optimal (3·9335 seconds). Weighted-sum GA using weights of 0·1, 0·0 and 0·9 managed to obtain the best motion time of 3·9531 seconds. However, this result actually considers only two objective functions, and cannot be used to compare with the results obtained using Pareto-based GA. In this case, the best motion time for normalised weight is 3·9570 seconds, obtained using weights of 0·3, 0·3 and 0·4 for w_1, w_2 and w_3 respectively.

The effect of the different number of generations is indicated in Figure 8.10, with Table 8.15 showing the effect of changing the size of the genetic population.

Table 8.15   Effects of different population size*

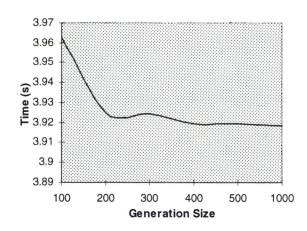

Figure 8.10   Generation size versus motion time

Population size	Minimum time(s)
100	4·0149
200	3·9804
300	3·9630

* Results obtained are using weights of (0·1,0·1,0·8) and 100 generation size.

(b) Non-normalised formulation

A set of non-normalised weights is also simulated. The results obtained when considering different combinations of weights are shown in Table 8.16, and show no motion time improvement in comparison with the normalised approach.

(c) Choice of weights

Moving from one set of weights to another may result in skipping a nondominated point. In other words, it is quite possible to miss using weights that would lead to extreme points (optima). Consequently, the most that should be expected from the weighting method is an approximation of the nondominated set. Although this approach may yield meaningful results only when solved many times for different values of weights, the results reported in this case show the difficulty of realising a solution. Normalised formulation has narrowed down the scope for searching for the appropriate combination of weights, but the range of joint travel times is only an approximation.

Table 8.16   Results for non-normalised formulation*

Weights($\omega_1,\omega_2,\omega_3$)	Minimum time(sec)
(400,200,400)	3·9984
(0·01,30,4000)	4·0057
(0·001,0·008,0·001)	4·0075
(5,4,1)	4·0260
(5,4,20)	3·9881
(5,1,20)	3·9441
(5,10,20)	4·0000
(5,0,20)	4·0066
(5,2,20)	3·9485
(4,1,20)	4·0209
(6,1,20)	3·9847
(8,1,20)	4·0034

* Results obtained using a population size of 200 and 100 generations.

8.3 Multiple manipulator systems

The problem of path planning of multiarm robots is different from that of single arm robots in that one arm may act as an obstacle to another arm. So the motion has to be planned for the two arms simultaneously. If the robots have to avoid collision with static obstacles in the workspace, and the additional constraints due to the restricted range in which the joint angles can be displaced are considered, the path planning problem becomes very complex for the conventional planners to solve [16,17]. GAs have been applied successfully to this problem, since they are able to handle different constraints by incorporating them in their fitness function. Although the formulation is given for two planner robots (see Figure 8.11), the simulation results are reported for two PUMA robots in 3D space, thus

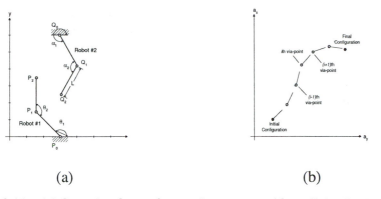

(a)   (b)

Figure 8.11   (a) *Operational space for two planner arms with two links, (b) via-points in configuration space*

indicating the generality of the algorithm.

8.3.1 Problem formulation

Path planning is carried out in configuration space of the manipulator. The entire path of the robot is considered simultaneously as a string of via points $\{p_0, p_1, ..., p_i, ..., p_N\}$ joining the initial configuration p_0 and the final configuration p_N, where p_i is the ith via point given by the ordered pair $(\theta_{1i}, \theta_{2i})$ and N is the total number of via points on the path. These via points are then fitted with parametric cubic splines. Repeated modification is carried out to the position of the via points through an evolutionary search to find a collision free path.

   In order to determine collision between the manipulator and the static obstacles, both the manipulator and the obstacles are approximated with touching circles. The distance between the centres of these circles and the static obstacles is checked. If this distance happens to be less than the sum of the radii of the circles, the manipulator is colliding with the obstacle; otherwise, it lies in the free space. An alternative to this is to use neural networks for collision detection [22]. The input to the neural networks is the joint angles of the manipulator, and the output varies between 0 and 1, depending upon whether the manipulator lies in free space or is colliding with the obstacle.

8.3.2 Encoding of paths as strings

For N via points, the paths are encoded directly as chromosomes of the evolutionary program as:

$$\mathbf{p}_1 \vdots \mathbf{p}_2 \vdots \cdots \mathbf{p}_{i-1} \vdots \mathbf{p}_i \vdots \mathbf{p}_{i+1} \vdots \cdots \mathbf{p}_{N-1}$$

where $\mathbf{p}_i$ is the vector forming the ith gene in the chromosome, and represents the ith via point on the path in the configuration space of the manipulator, and $\vdots$ is the concatenation operator. Thus, each gene in the chromosome consists of a vector with floating point components.

8.3.3 Fitness function

Different objectives to be minimised by the evolutionary algorithm are:

$C_1 =$ Penalty on the length of the path in configuration space given by

$$C_1 = \sum_{i=1}^{N}\sqrt{(\theta_{1i}-\theta_{1(i-1)})^2 + (\theta_{2i}-\theta_{2(i-1)})^2},$$ where N is the total number of via points and $\mathbf{p}_i$ $(\theta_{1i}, \theta_{2i})$ is the ith via point on the path of the manipulator.

$C_2 =$ Penalty on the uneven distribution of via points on the path defined as

$$C_2 = \sum_{i=1}^{N}\|d_i - d\|,$$ where $d_i =$ Euclidean distance between ith and

$(i-1)$th via point, and $d = \frac{1}{N}\sum_{i=1}^{N}d_i.$

$C_3 =$ Penalty on collision with the obstacles,

$$C_3 = \max_{i}(K_i); \; i = 1,2,..,(N-1),$$

where $K_i = \begin{cases} 1 & \text{if the manipulator collides with the obstacle in the } i\text{th configuration} \\ 0 & \text{otherwise} \end{cases}$

This latter function is referred to as a hard-threshold collision function since it only tells whether the manipulator is colliding with the obstacle or not, and does not give any information as to how far the manipulator is from the obstacle. The value of the function steps from 0 to 1 when the manipulator moves from free space into the *c*-obstacle without a gradual slope.

Thus, the fitness function is formulated in two ways:

(i)   Linear combination of objectives.
(ii) Prioritisation of objectives.

(*i*) *Linear combination of objectives*
The objective function to be maximised is given by a linear combination of objectives as:

$$g = k_1 \frac{C_1}{C_{1\ average}} + k_2 \frac{C_2}{C_{2\ average}} + k_3 \frac{C_3}{C_{3\ average}}$$

where k_1, k_2 and k_3 are positive constants and $C_{1\ average}$, $C_{2\ average}$ and $C_{3\ average}$ are the averages of the penalties on path length, the uneven distribution of via points on the path and collision with the obstacles, respectively, calculated for the initial population at the beginning of the search. The fitness function is again formulated in two ways. In the first case, the fitness function is formulated as:

$$F = \begin{cases} C_{max} - g & \text{if } g < C_{max} \\ 0 & \text{otherwise} \end{cases}$$

where C_{max} is a positive constant. It may be pointed out that if the hard threshold collision detection is chosen, then the maximum value of C_2 would be 1, so the value of $C_{2\ average}$ is taken to be equal to 1 in order to give it the same weight as the other objectives. The normalisation of all the penalties by dividing them by their respective initial averages provides a way of visualising the weight associated with each penalty easily, thus making it easy to choose the values of k_1, k_2 and k_3. The value of k_3 is kept relatively high, since collisions are to be avoided at all costs. The value of k_2, on the other hand, is kept very low, since the via points are redistributed evenly by one of the operators (the redistribution operator) in the algorithm. The value of C_{max} is chosen so that it is higher than the expected value of *g* at all times.

One problem associated with selection based on this type of fitness function is that a string with a relatively high fitness could fill up the entire population very quickly, thus resulting in a premature convergence of the algorithm to a suboptimal solution. To counter this problem, a ranking of the population has been investigated, in which the strings in the population are not chosen according to their fitness value, but according to their rank among the population. The rank of an individual depends upon its fitness in a descending order in the population. A linear ranking has been used, in which the rank varies from a maximum value to a minimum value.

(*ii*) *Prioritisation of objectives*

Among the three objectives defined, the collision avoidance is a constraint, whereas the other two objectives, i.e. penalty on the path length and penalty on the uneven distribution of via points, are objectives to be minimised. One way of handling these two different types of objective is to formulate a fitness function depending on the priority of the objective. Highest priority is assigned to the collision avoidance, and minimisation is performed first on this objective. The other objectives are minimised at a lower priority. The fitness function is defined as:

$$ g = \max\left(\left\| k_1\left(\frac{C_1 - C_{st}}{C_{1average} - C_{st}} \right) + k_2\left(\frac{C_2}{C_{2average}} \right) \right\|, \left\| k_3\left(\frac{C_3}{C_{3average}} \right) \right\| \right) $$

where C_{st} is the length of straight line path in configuration space between the initial configuration and the final configuration, and the fitness function F is defined as above. The value of k_3 is kept higher than that of both k_2 and k_1 to give it a higher priority.

8.3.4 The GA operators

The following operators are used in the evolutionary algorithm:

(a) Reproduction: the strings are reproduced for the next generation based on their fitness function. A weighted roulette wheel is used to select the strings from an old population for a new population.
(b) Crossover: the individuals in the population reproduced from the old population based on their fitness are grouped at random into pairs of parent strings. Some crossover site is chosen at random among two parent strings. A cross-over is then performed by switching position of the via points between this site and the two parent strings to produce two offspring strings. This operation is carried out with a certain probability and only if the distance between the crossover sites is less than a certain value.

(c) Redistribution: the via points are fitted with parametric cubic splines and then redistributed over these splines at equal distance to make the distribution even.

(d) Relaxation: the path is then made to behave like a stretched string and relax under the strain (see Figure 8.12).

(e) Mutation: in order to carry out the mutation (which is done at a probability of *mutation_probability*) any gene (via point) in the chromosome is selected, and random values within a specific range are added to all the components in the gene.

(f) Regeneration: new trajectories are generated and are injected into the population after every generation.

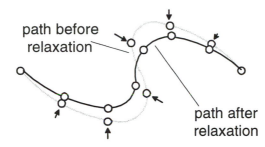

Figure 8.12   Effect of relaxation on the path, indicating the change in the position of via points on the path

8.3.5 Simulation results for two 3DOF arms

The parameter values given in Table 8.17 were used for the algorithm, and the initial and final configurations for the robots are given in Table 8.18. In addition, the bounds on the displacement of the joint angles are given in Table 8.19. Figure 8.13 shows the motion of the two arms in operational space and Figure 8.14 show the history of the fitness function and the paths in operational space, respectively.

Table 8.17   Values of parameters used in the evolutionary search algorithm for two 3DOF arms moving in 3D space with static obstacles in the environment

k_1	200
k_2	50
k_3	200
k_4	200
k_5	50
C_{max}	1000
population size	100
new_trajectories	10
keep_best	1
cross_over_distance	100°
cross_over_probability	1.0
mutation_probability	0.001

Table 8.18   Initial and final configurations for two 3DOF manipulators moving in 3D operational space

	Initial value	Final value
θ_1	15°	150°
θ_2	-15°	-15°
θ_3	-30°	30°
α_1	-150°	-15°
α_2	-15°	-15°
α_3	30°	-30°

Table 8.19   Lower and upper bounds on the joint angle displacements for the two 3-DOF manipulators

	Lower bound	Upper bound
θ_1	0°	180°
α_1	-180°	0°
θ_2 and α_2	-30°	30°
θ_3 and α_3	-45°	45°

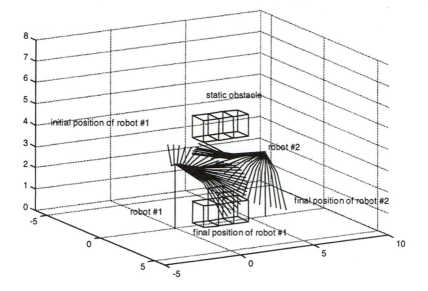

Figure 8.13   Collision free motion of two 3DOF arms in 3D operational space

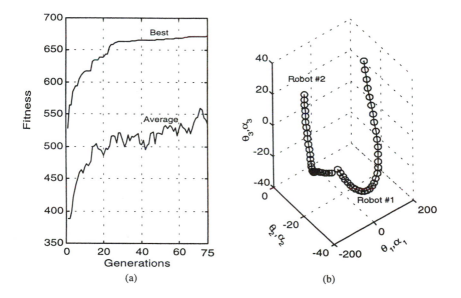

Figure 8.14   (a) History of the fitness function and (b) paths in the configuration space of two 3DOF manipulators

8.4 Mobile manipulator system with nonholonomic constraints

A mobile platform with an onboard manipulator, as shown in Figures 8.15 and 8.16, is considered in this Section. The manipulator has one rotational link and two planar links. The platform has two driving wheels (the centre ones) and four passive wheels (the corner ones). The two driving wheels are independently driven by two motors.

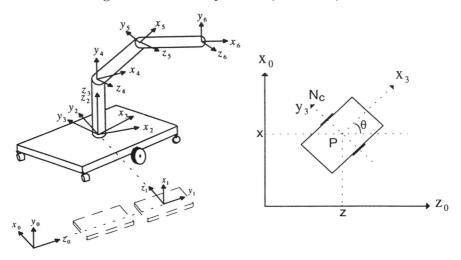

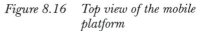

Figure 8.15   A mobile platform with an   Figure 8.16   Top view of the mobile
onboard manipulator   platform

The kinematic relations for a mobile manipulator can be represented as:

$$\mathbf{X}_e(t) = \mathbf{X}_p(t) + \mathbf{X}_{m/p}\,(\Phi(t))$$

where $\mathbf{X}_e(t)$ is the given Cartesian trajectory of the end effector in the world frame, $\mathbf{X}_p(t)$ is the Cartesian position trajectory of the platform in the world frame, $\mathbf{X}_{m/p}\,(\Phi(t))$ represents the vector of the position of the end effector with respect to the platform reference frame and Φ is the vector of manipulator joint angles, which are usually constrained by independent upper and lower bounds. The platform is subject to the no slipping nonholonomic constraint

$$-\,\dot{z}\sin\theta + \dot{x}\cos\theta = 0$$

i.e., the platform must move in the direction of the axis of symmetry. Note that the position (z, x) and the heading angle θ of the platform are not independent of each other due to the nonholonomic constraint.

8.4.1 Multicriteria cost function

Various optimisation criteria can be applied in the trajectory generation of the mobile manipulator system. In this study, total actuator torque minimisation and maximising the manipulator's manipulability measure are considered. The total actuator torque over the time interval starting at $t = t_0$ and ending at $t = t_f$ is defined as:

$$f_1 = \int_{t_0}^{t_f} \tau^T \tau \, dt$$

where τ is the actuator torque vector and is constrained by independent upper and lower bounds. The manipulability measure can be regarded as a distance measure of the manipulator configuration from singular configurations at which the manipulability measure becomes zero. At or near a singular configuration, the end point of the manipulator may not easily move in certain directions. Maximising manipulability keeps the manipulator away from singular points and provides more velocity transmission ratio in all directions. The manipulator manipulability measure is defined as [14]:

$$\omega = \sqrt{\det(J(\Phi)J^T(\Phi))}$$

where $J(\Phi)$ is the manipulator Jacobian matrix and Φ is the vector of manipulator joint angles. The total cost function is thus defined as:

$$f = \varepsilon \int_{t_0}^{t_f} \tau^T \tau \, dt + \lambda \int_{t_0}^{t_f} (\omega_m - \omega) \, dt$$

where ω_m is a given positive real number not less than the possible maximum manipulability, ε and λ are the relative weightings between the two criteria, which are under the control of users. For obstacle avoidance, a geometric analysis method [15] is applied. For simplicity, the boundaries of the platform, the manipulator and the obstacles are represented by a number of straight lines. If any boundary line of the robot intersects with any boundary line of the obstacles, collision occurs and the robot path will be modified.

The optimal trajectory-generating problem is thus stated as follows: given the trajectory of the end effector of the manipulator, search for near optimal trajectories for the platform and the manipulator joints to minimise the total actuator torque and maximise the manipulability with a nonholonomic constraint and obstacle avoidance.

8.4.2 Parameter encoding using polynomials

Generally it is desirable for a robot joint or the platform to move along a smooth trajectory because smooth motion needs less energy and avoids structural resonance. Thus, the initial joint and platform trajectories are represented by polynomials which are continuous and have continuous first derivatives. Suppose the initial and final positions and velocities of a trajectory are given, then at least a three order polynomial is needed to represent it. In order to have some free parameters which can be chosen to optimise the trajectory, a higher order polynomial is chosen, namely a fifth order. Suppose a generalised joint trajectory is represented by a fifth order polynomial as follows:

$$\theta(t) = \alpha_0 + \alpha_1 t + \alpha_2 t^2 + \alpha_3 t^3 + \alpha_4 t^4 + \alpha_5 t^5, \quad t_0 \leq t \leq t_f$$

The angles and velocities at the beginning and the end of a trajectory are given as: $\theta_0 = \theta(t_0)$, $\theta_f = \theta(t_f)$, $\dot{\theta}_0 = \dot{\theta}(t_0) = 0$, $\dot{\theta}_f = \dot{\theta}(t_f) = 0$. Two additional constraints are required to determine the parameters of the above polynomial. Here the initial and ending accelerations of a joint trajectory are introduced as the additional constraints required. The $2n_r$ initial and ending accelerations of the n_r chosen joint trajectories are encoded in the genetic algorithm. The binary genetic string is thus generated as follows:

$$\{a_{10}, a_{1f}, ..., a_{i0}, a_{if}, ..., a_{n_r0}, a_{n_rf}\}$$

where a_{i0} and a_{if} are the initial and ending accelerations of the ith chosen generalised joint trajectory. Here, the n_r generalised joint trajectories for encoding can be chosen randomly. Then, the polynomial trajectory of each encoded generalised joint can be determined by the generated initial and ending accelerations combined with the angle and velocity constraints at the beginning and the end of each trajectory.

After the trajectories of the encoded joints have been determined, the remaining generalised joint trajectories can be calculated in the following way. Discretise the time interval starting at $t = 0$ and ending at $t = t_f$ into $N+1$ equal sections. Beginning with the starting position, the joint value at each via point of each remaining trajectory is calculated from the end effector's Cartesian trajectory by using the inverse kinematics combined with the nonholonomic constraints. The joint values at the N via points can be represented as:

$$\{ \theta_{i,1}, \theta_{i,2}..., \theta_{i,j-1}, \theta_{i,j}, ...\theta_{i,N} \}$$

where $\theta_{i,j}$ is the joint value of the ith joint at the jth via point and i represents one of the remaining joints which is not encoded. At the same time, the joint value at every via point is checked to see whether it meets the joint lower and upper bound. In the case where there are obstacles in the working space, the geometric method is applied to check whether, in any position, the platform trajectory collides with the obstacles. If at any via point any trajectory violates the kinematic constraints or collides with obstacles, the joint value of this trajectory at this via point is discarded and a new joint value is regenerated as follows until a valid value is produced:

$$\theta_{i,j} = \theta_{i,j-1} + r_{i,j} \Delta_i$$

where $\theta_{i,j}$ is the randomly produced new joint value at point j, and $\theta_{i,j-1}$ is the joint value at point j-1, Δ_i is a small given positive number and $r_{i,j}$ is a randomly generated integer from a known range.

8.4.3 Fitness function

For this problem, the fitness function is defined as:

$$F = f_{max} - f - f_p$$

where f is the cost function defined in Section 4.1, f_{max} is a properly selected positive real number not less than the maximum value of f, and f_p is a torque penalty which is $f_{max}/20$ when any trajectory violates the upper or the lower torque bounds, and otherwise is set to 0.

8.4.4 Genetic evolution

A reproduction approach is applied to the selection of strings for the next generation. In order to reduce the stochastic error associated with the selection, the stochastic remainder sampling scheme without replacement is applied. Other strategies applied include fitness scaling and random migration.

The crossover operation is applied as follows. Members of the newly reproduced strings are paired at random. For each pair of selected strings, with a probability of p_c, a cross position is selected at random. The two new strings are created by swapping parts of the strings from the selected position to the last position. The result is that two new sets of initial and ending accelerations are formed, that is, two new sets of joint trajectories are produced.

The mutation operator changes individual strings on a bit by bit basis, with a very small probability p_m. Once a mutation is performed in a string, a new set of boundary accelerations are formed. The main purpose of mutation is to bring in new information and to protect against loss of some potentially useful genetic material gained during reproduction and crossover.

The fitness values of the new population's strings are evaluated, and the process continues until a predefined number of generations is reached. Since all initial joint trajectories are represented as polynomials, smooth trajectories are produced and a low cost for each trajectory is expected. Therefore, the genetic algorithm requires a smaller population and converges to a near optimum in fewer generations than by via point representation.

8.4.5 Simulation results

A case study of a system with a PUMA-like three-link manipulator mounted on a mobile platform is considered. One typical set of genetic parameters used in testing the system is: population size $n = 50$, generation number $g = 40$, crossover probability $p_c = 0.8$, mutation probability $p_m = 0.03$ and the number of bits for each real acceleration in the genetic string is 16. The cost function f (of Section 8.4.1) was used in evaluating the fitness with weightings $\varepsilon = 1$ and $\lambda = 1$. All the simulations were conducted on a Sun Sparc station and the computation time was less than 100 seconds.

Two simulations are reported, one without obstacles (case 1) and another with a rectangular obstacle placed in the workspace of the platform (case 2), with the two resulting motions shown in Figures 8.17 and 8.18, respectively.

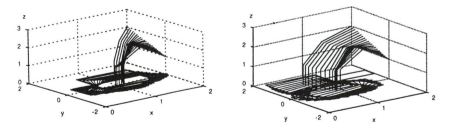

Figure 8.17   *Case 1 – motion without an obstacle*   *Figure 8.18*   *Case 2 – motion with obstacle avoidance*

The results for case 1 are: total manipulability measure is 22·532 (ω_m is chosen as 30), total torque is 16·106 and the total cost is 23·574. For case 2, the manipulability measure is 24·733 and the total torque is

20·131 with a total cost of 25·397. Although the manipulability is better in case 2 than in case 1 due to obstacle avoidance, the total cost is higher in case 2. The manipulability is shown in Figure 8.19, with the planned trajectories in Figure 8.20, both for case 2 of the simulations.

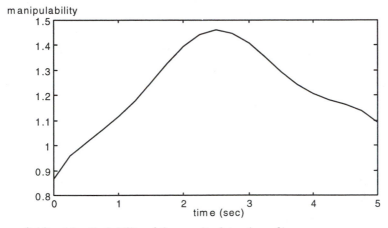

Figure 8.19   Manipulability of the manipulator (case 2)

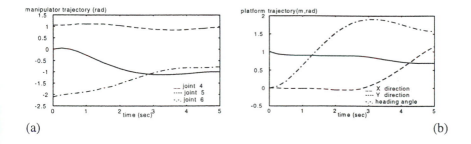

Figure 8.20   (a) Platform trajectories, (b) manipulator trajectories (case 2)

8.5 Discussions and conclusions

The search for minimum-time motion for an articulated mechanical arm by tessellating the joint space involves a heavy computational burden. In this work, evolutionary algorithms are used to tackle this problem. A genetically-based motion planner is formulated for a six-joint articulated arm considering all physical constraints. In addition, different optimisation objectives were considered, namely minimum motion time, constrained motor torque values and constrained end point velocities. The applicability and efficiency of the proposed algorithm are demonstrated via several simulations on the PUMA manipulator. Further work on the planner has allowed investigations into initial population assembly and trajectory decoding mechanisms, thus leading to much enhanced performance. The algorithm was also implemented using a migration-type parallel GA with the population distributed into three small subgroups. Each subgroup contributes the best chromosome to the root processor, thus outperforming the sequential planning algorithm.

In comparing the Pareto GA, breeder GA, weighted-sum GA and the flexible polyhedron search methods, based on comprehensive simulations for the RTX robot, the following points were observed.

1 The customised path redistribution-relaxation operator has produced the best performance in the case studies among all tested operators, supporting the idea of custom-building operators to the requirements of an application.
2 Breeder GA operators can perform better in a nonBGA environment (e.g. truncation selection environment).
3 Results produced by the weighted-sum GA depend on the weight given to the objective function, generation size and population size. Hence there are too many parameters to be tested to obtain an optimum solution. The difficulties with setting these initialisation parameters are high and time-consuming.
4 Both GA methods, weighted-sum and Pareto-based, require a significant amount of processing time to obtain the optimum motion time, although FPS requires much less computation.
5 Pareto-based GA appears to have an edge in optimisation over weighted-sum GA, because it can produce a better optimum motion time (although small in comparison) using smaller generation and population sizes.
6 The optimality of the motion time is improving with population and generation size. However, a compromise is needed to ensure limiting processing time for complex cases.

An offline approach to time optimal path planning of multiarm robotic systems based on evolutionary programming has been presented. The path is represented by a string of via points between the initial and final configuration in joint space, with the trajectories between these via points interpolated by cubic splines. Repeated path modification by changing the position of the via points is carried out by the evolutionary algorithm to search for a global near-time optimal solution. The strength of this algorithm lies in the fact that the planning can be extended to include additional constraints (such as energy minimisation) by incorporating them into the fitness function. In addition, unlike variational calculus techniques, the algorithm does not get stuck at local minima. The proposed algorithm also suits time-optimal motion planning of redundant manipulator systems, where collisions between links can be avoided by considering them as moving obstacles. Moreover, since only forward kinematics is used in collision avoidance, no singularities are encountered, which are inherent in inverse kinematics of redundant manipulators. The algorithm has been extended to include static obstacles in the workspace, and its application to practical systems is being investigated. An interesting area which needs to be explored is determining how close the bounds on torque values can be set to the actual saturation limit so as to provide enough play for the controller to account for the uncertainties in the model of the practical systems.

The optimal trajectory generation problem for a mobile manipulator system with nonholonomic constraints and obstacle avoidance is a nonlinear multicriteria optimisation problem, where the solution space is discontinuous and contains local minima. A robust genetic algorithm is applied to solving this problem with torque minimisation, manipulability maximisation and obstacle avoidance. More criteria can easily be combined into the cost function. Computational efficiency of the genetic algorithm is achieved by applying a polynomial method. Although only simulations for a mobile 3DOF robot are reported here, the algorithm has been successfully implemented on Sheffield's LongArm system comprising a B12 mobile and a 2DOF arm.

Further details on the reported algorithms can be found in the literature [18-25], or by accessing the group's web site on http://www.shef.ac.uk/rrg.

8.6 Acknowledgment

Parts of the reported work were supported by the Engineering and Physical Sciences Research Council (grant GR/J15797).

8.7 References

1   Kahn, M. E., and Roth B.: 'The near-minimum-time control of open-loop articulated kinematic chains'. AIM 106. Stanford: Stanford Artificial Intelligence Laboratory, 1971

2   Lin, C.-S., Chang, P.-R. and Luh, J. Y. S.: 'Formulation and optimisation of cubic polynomial trajectories for industrial robots', *IEEE Trans.*, **AC-28**, (12), pp. 1066–1073, 1983

3   Sahar, G., and Hollerbach, J. M.: 'Planning of minimum-time trajectory for robot arms', *Int. J. Robotics Res.*, **5**, (3) pp. 91–100, 1986

4   Micalewicz: *Genetic algorithms + data structures = evolution programs* (Springer-Verlag, 1992)

5   Chipperfield, A. J., and Fleming P. J.: 'Parallel genetic algorithms: a survey'. Research report, Univ. of Sheffield, May 24, 1994

6   Dissanayake, M. W. M. G., Goh, G. J., and Phan-Thien, N.: 'Time-optimal trajectories for robot manipulators', *Robotica*, **9**, pp. 131–138, 1991

7   Fonseca, C. M., and Fleming, P. J.: 'An overview of evolutionary algorithms in multiobjective optimisation', *Evolutionary Computation.*, **3**, (1), pp. 1–16, 1995

8   Goldberg, D. E.: *Genetic algorithms in search, optimisation & machine learning* (Reading: Addison-Wesley Publishing Company, Inc., 1989)

9   Himmelblau, D. M.: *Applied non-linear programming* (New York: McGraw-Hill, Inc., 1972)

10 Whitley, D.: 'The GENITOR algorithm and selection pressure: why rank-based allocation of reproductive trials is best', in proceedings of the third international conference on *Genetic algorithms*, Schaffer, J. D. (ed.) (Morgan Kaufmann, 1989), pp. 116–121

11 Baker, J. E.: 'Reducing bias and inefficiency in the selection algorithm', in *Proceedings of the second international conference on genetic algorithms*, Grefenstette, J. J. (ed.) (Lawrence Erlbaum Associates, Publishers, 1987) pp. 14–21

12 Mühlenbein, H., and Schlierkamp-Voosen, D.: 'Predictive models for the breeder genetic algorithm, I: continuous parameter optimisation', *Evolutionary Computation*, **1**, (1), pp. 25–49, 1993

13 Rana, A.S., and Zalzala, A.M.S.: 'Near time-optimal collision-free motion planning of robotic manipulators using an evolutionary algorithm', *Robotica*, **14**, pp. 621–32, 1996

14 Yamamoto, Y., and Yun, X.: 'Co-ordinating locomotion and manipulation of a mobile manipulator', *IEEE Trans. Autom. Control*, **39**, (6), June, pp. 1326–32, 1994

15 Bowyer, A., and Woodwark, J.: *A programmer's geometry* (Butterworths, 1983)

16 Hwang, Y. K., and Ahuja, N.: 'Gross motion planning – a survey', *ACM Comput. Surv.*, **24**, (3) Sept. 1992

17 Latombe, J. C.: *Robot Motion Planning* (Kluwer Academic Publishers, Amsterdam, 1991)

18 Chen, M., and Zalzala, A.M.S.: 'Dynamic modelling and genetic-based motion planning of mobile manipulator systems with nonholonomic constraints', *Control Eng. Pract.*, **5**, (1), pp. 39–48, 1997

19 Chen, M., and Zalzala, A.M.S.: 'A genetic approach to the motion planning of redundant mobile manipulator systems considering safety and configuration', *J. Robot. Syst.*, 1997

20 Wang, Q., and Zalzala, A.M.S.: 'Genetic algorithms for PUMA robot motion control: a practical implementation', *Int. J. Mechatronics*, **6**, (3) pp. 349–65, 1996

21 Zalzala, A.M.S., and Fleming, P.J.: 'Genetic algorithms: principles & applications in engineering systems', *Int. J. Neural Network World*, **6**, pp. 803–20, 1996

22 Rana, A.S., and Zalzala, A.M.S.: 'A neural network based collision detection engine for multi-arm robotic systems', Proc. 5th Int. Conf. *Articifical Neural Networks*, IEE Conference 440, 1997, pp.140–45

23 Rana, A.S., and Zalzala, A.M.S.: 'An evolutionary planner for near time-optimal collision-free motion of multi-arm robotic manipulators', Proc. UKACC int. conf. on *Control*, **1**, pp. 29–35, Exeter, 1996

24 Wang, Q., and Zalzala, A.M.S.: 'Investigations into robotic multi-joint motion considering multi-criteria optimisation using genetic algorithms'. Proc. *IFAC* world congress, **A**, pp. 301–6, San Fransisco, 1996

25 Wang, Q., and Zalzala, A.M.S., 'Investigations into the decoding of genetic based robot motion considering sequential and parallel formulations'. Proc. UKACC int. conf. on *Control*, **1**, pp. 442–7, Exeter, 1996

8.8 Appendix

8.8.1 Motion based on cubic splines

The objective here is to construct joint trajectories which fit a number of joint displacements at a sequence of time instants by using cubic polynomial functions. Consider a vector of via points for the jth joint along some initial path as $[\theta_{j1}(t_1), \theta_{j2}(t_2), ..., \theta_{jn}(t_n)]$, where $t_1 < t_2 < t_3 < t_4 <$

$\dots < t_{n-2} < t_{n-1} < t_n$ is an ordered time sequence, indicating that the position of the jth joint at time $t = t_i$ is $\theta_{ji}(t_i)$. Let v_{ji} and w_{ji} denote the velocity and acceleration of joint j at knot i. At the initial time $t = t_1$ and the terminal time $t = t_n$, the joint displacement, joint velocity and joint acceleration are θ_{j1}, v_{j1}, w_{j1} and θ_{jn}, v_{jn}, w_{jn} respectively. In addition, joint positions θ_{jk} at $t = t_k$ for $k = 3, 4,\dots, n-2$ are also specified, However, θ_{j2} and, $\theta_{j,n-1}$ are the two extra knots required to provide the freedom for solving the cubic polynomials. For simplicity, the subscript j for jth joint is dropped since the result is the same for all the joints. From Reference 6, the acceleration vector w of knots is obtained as follows:

$$A\,w = b \tag{8.1}$$

where: $w = [\,w_2,\ w_3, w_4, \dots, w_{n-3},\ w_{n-2},\ w_{n-1}\,]^T$

$$A = \begin{bmatrix} 3h_1 + 2h_2 + \dfrac{h_1^2}{h_2} & h_2 & \cdots & 0 & 0 \\[2ex] h_2 - \dfrac{h_1^2}{h_2} & 2(h_2 + h_3) & \cdots & 0 & 0 \\[2ex] 0 & h_3 & \cdots & 0 & 0 \\[1ex] \cdots & & \cdots & & \\[1ex] 0 & 0 & \cdots & 2(h_{n-3} + h_{n-2}) & h_{n-2} - \dfrac{h_{n-1}^2}{h_{n-2}} \\[2ex] 0 & 0 & \cdots & h_{n-2} & 3h_{n-1} - 2h_{n-2} + \dfrac{h_{n-1}^2}{h_{n-2}} \end{bmatrix},$$

$$b = \begin{bmatrix} 6\left(\dfrac{\theta_3}{h_2} + \dfrac{\theta_1}{h_1}\right) - 6\left(\dfrac{1}{h_1} + \dfrac{1}{h_2}\right)\left(\theta_1 + h_1 v_1 + \dfrac{h_1^2}{3} w_1\right) - h_1 w_1 \\[2ex] \dfrac{6}{h_2}\left(\theta_1 + h_1 v_1 + \dfrac{h_1^2}{3} w_1\right) + \dfrac{6\theta_4}{h_3} - 6\left(\dfrac{1}{h_2} + \dfrac{1}{h_3}\right)\theta_3 \\[2ex] 6\left(\dfrac{\theta_5 - \theta_4}{h_4} - \dfrac{\theta_4 - \theta_3}{h_3}\right) \\[1ex] \cdots \\[1ex] \dfrac{6}{h_{n-2}}\left(\theta_n - v_n h_{n-1} + \dfrac{h_{n-1}^2}{3} w_n\right) - 6\left(\dfrac{1}{h_{n-2}} + \dfrac{1}{h_{n-3}}\right)\theta_{n-2} + \dfrac{6}{h_{n-3}}\theta_{n-3} \\[2ex] -6\left(\dfrac{1}{h_{n-1}} + \dfrac{1}{h_{n-2}}\right)\left(\theta_n - v_n h_{n-1} + \dfrac{h_{n-1}^2}{3} w_n\right) + \dfrac{6\theta_n}{h_{n-1}} + \dfrac{6\theta_{n-2}}{h_{n-2}} - h_{n-1} w_n \end{bmatrix}$$

and

$$h_i = t_{i+1} - t_i \ (i = 1, 2, ..., n-1) \tag{8.2}$$

$$w_i(t) = \frac{t_{i+1} - t}{h_i} w_i + \frac{t - t_i}{h_i} w_{i+1} \tag{8.3}$$

$$v_i(t) = -\frac{w_i}{2h_i}(t_{i+1} - t)^2 + \frac{w_{i+1}}{2h_i}(t - t_i)^2 + \left(\frac{\theta_{i+1}}{h_i} - \frac{h_i w_{i+1}}{6}\right) - \left(\frac{\theta_i}{h_i} - \frac{h_i w_i}{6}\right) \tag{8.4}$$

$$\theta_i(t) = \frac{w_i}{6h_i}(t_{i+1} - t)^3 + \frac{w_{i+1}}{6h_i}(t - t_i)^3 + \left(\frac{\theta_{i+1}}{h_i} - \frac{h_i w_{i+1}}{6}\right)(t - t_i)$$

$$+ \left(\frac{\theta_i}{h_i} - \frac{h_i w_i}{6}\right)(t_{i+1} - t) \tag{8.5}$$

where $i = 1, ..., n-1$. The positions, velocities and accelerations can therefore be obtained, provided that each time interval h_i is known.

The optimal time solution for the time interval vector X should be obtained with the corresponding knot velocities and accelerations computed from the above equations, and the jerks are the rate of change for the corresponding accelerations. The resulting via point velocities, accelerations and jerks may or may not violate their limits and should be compared with their own limits to obtain the time-optimal path satisfying the joint constraints.

8.8.2 Physical limits

Note that there are six joints that must be considered simultaneously and there are three constraints, i.e. velocity, acceleration and jerk limits for each joint. For convenience, let:

VC_j	= velocity constraint for joint j
WC_j	= acceleration constraint for joint j
JC_j	= jerk constraint for joint j
$Q'_{ji}(t)$	= velocity for joint j between knot i and $i+1$
$Q''_{ji}(t)$	= acceleration for joint j between knot i and $i+1$
$Q'''_{ji}(t)$	= jerk for joint j between knot i and $i+1$
X	= $(h_1, h_2, ..., h_{n-1})$, the vector of time intervals
$Q_{ji}(t)$	= piece wise cubic polynomial trajectory for joint j between knot i and $i+1$
w_{ji}	= acceleration of joint j at knot i

The acceleration in eqn. 8 3 can therefore be rewritten as:

and also the velocity function in eqn. 8.4 can be rewritten as:

$$Q''_{ji}(t) = \frac{t_{i+1} - t}{h_i} w_{ji} + \frac{t - t_i}{h_i} w_{j,i+1}$$

$$Q'_{ji}(t) = -\frac{w_{ji}}{2h_i}(t_{i+1} - t)^2 + \frac{w_{j,i+1}}{2h_i}(t - t_i)^2 + \left(\frac{\theta_{j,i+1}}{h_i} - \frac{h_i w_{j,i+1}}{6}\right) - \left(\frac{\theta_{ji}}{h_i} - \frac{h_i w_{ji}}{6}\right)$$

The objective thus is to minimise $\sum\limits_{i=1}^{n-1} h_i$ subject to constraints:

$$|Q'_{ji}(t)| \le VC_j, \quad |Q''_{ji}(t)| \le WC_j \text{ and } |Q'''_{ji}(t)| \le JC_j$$

for $j = 1, 2, ..., N$ and $i = 1, 2, ..., n-1$.

8.8.3 The feasible solution converter (time scaling)

If the correct time intervals are guessed, then w_{ji} can be uniquely determined from eqn. 8.1. However, if constraints on joint velocities, accelerations and jerks are not satisfied, then time intervals $[h_1, h_2, ..., h_{n-1}]$ should be expanded to bring the unsatisfied velocities, accelerations, and jerks to their constrained values. Now let:

$$\lambda_1 = \max_j \left[\max_{t \in [t_i, t_{i+1}] \forall i} |Q'_{ji}(t)| / VC_j \right]$$

$$\lambda_2 = \max_j \left[\max_{t \in [t_i, t_{i+1}] \forall i} |Q''_{ji}(t)| / WC_j \right] = \max_j \left[\max_i |w_{ji}| / WC_j \right]$$

$$\lambda_3 = \max_j \left[\max_{t \in [t_i, t_{i+1}] \forall i} |Q'''_{ji}(t)| / JC_j \right] = \max_j \left[\max_i \left| \frac{w_{j,i+1} - w_{ji}}{h_i} \right| / JC_j \right]$$

$$\lambda = \max\left(1, \lambda_1, \sqrt[2]{\lambda_2}, \sqrt[3]{\lambda_3}\right).$$

If the time interval h_i is replaced by λh_i for $i = 1, 2, ... \; n-1$, then the velocity, aceleration and jerk will be replaced by factors of $\frac{1}{\lambda}, \frac{1}{\lambda^2}, \frac{1}{\lambda^3}$, respectively. These changes assure the satisfaction of constraints on velocities, accelerations and jerks.

Chapter 9
Aerodynamic inverse optimisation problems

S. Obayashi

Development of an aerodynamic shape optimisation method is important for the commercial aircraft industry to improve the design efficiency in today's competitive environment. With the aid of computational fluid dynamics (CFD), various aerodynamic design techniques have been proposed. CFD codes compute the flows around aircraft, and thus designers of an aircraft can predict its aerodynamic performance from the flow solution. Existing CFD codes have been coupled with various optimisation algorithms to obtain better aerodynamic design.

Among the numerical optimisation algorithms, gradient-based methods have been used widely. The optimum obtained from these methods will be a global optimum, if the objective and constraints are differentiable and convex [1]. In practice, however, it is very difficult to prove differentiability and convexness. One can only hope for a local optimum in a neighbour of the initial point, provided that the gradient is well defined. Therefore, one must start the design from various initial points to see if one can obtain a consistent optimum and therefore have reasonable assurance that this is the true optimum. In this sense, the gradient-based methods are not robust.

Evolutionary algorithms, in particular, genetic algorithms (GAs), are known to be robust optimisation algorithms [2] and have been enjoying increasing popularity in the field of numerical optimisation in recent years. GAs are search algorithms based on the mechanics of natural selection and natural genetics. One of the key features of GAs is that they search from a population of points, not a single point. In addition, they use objective function information (fitness value), not derivatives or other auxiliary knowledge. These features make GAs robust and thus attractive to practical engineering applications. GAs have been applied to aeronautical problems in several ways, including the parametric and conceptual design of aircraft [3,4], preliminary design of turbines [5], topological design of nonplanar wings [6] and aerodynamic optimisation using CFD [7-10].

With an appropriate choice of optimisation algorithm, aerodynamic numerical optimisation methods are categorised into two classes [11]: direct and inverse numerical optimisation methods. The direct numerical optimisation methods are formed by coupling aerodynamic analysis methods with numerical optimisation algorithms. They minimise (or maximise) a given aerodynamic objective function by iterating directly on the geometry. The geometry is represented by a general function, such as polynomial and cubic splines, by a linear combination of known airfoils, or by a basic shape plus a combination of typical geometry perturbations.

In this Chapter, the direct approach to airfoil shape optimisation is first considered to evaluate performance of the existing optimisation algorithms. One of these is the gradient-based method (GM). Another is simulated annealing (SA) [12]. SA is a heuristic strategy for obtaining near-optimal solutions, and derives its name from an analogy to the annealing of solids. The other is GA, one of the evolutionary algorithms. The design result demonstrates the superiority of GA for aerodynamic optimisation among the others.

Applicability of GAs to CFD problems, however, was limited by the fact that they were the direct numerical optimisation methods. CFD is a typical example of large-scale simulations. The direct approach requires CFD evaluation of each member of the population at every generation in GAs. As a result, it requires a tremendous amount of computational time. The inverse approach will therefore alleviate the computational time for engineering purposes.

The inverse numerical optimisation methods deal with pressure distributions rather than the geometry, to minimise, for example, drag under given lift and pitching moment. Since pressure is the primary force acting on aircraft, one can design desired aerodynamic characteristics by specifying pressure distributions. Once the target pressure distribution is optimised, corresponding geometry can be determined by the inverse methods.

Inverse methods themselves form a class of powerful design tools. These methods solve the classical inverse problem of determining the aerodynamic shape which will produce given pressure distributions. However, they leave the user with the problem of translating his design goals into properly defined pressure distributions exhibiting the required aerodynamic characteristics [13]. Although skilful designers are capable of producing successful designs, design efficiency can be improved by providing the designer with tools for target pressure specification. For this purpose, numerical optimisation of target pressure distributions has been studied in [14] and [15]. This approach avoids most of the limitations of the standard inverse methods and requires considerably less computational effort than the direct numerical optimisation approach. The main topic of this Chapter is the development of the inverse optimisation method using GA.

In the following, optimisation of wing shape is considered among aircraft components, since wing shape has the primary impact on the aircraft performance. The design of wings usually proceeds in two steps. First, the midspan section of the wing called the airfoil is designed. Since a typical wing for commercial aircraft has a longer span than its chord, wing performance can be predicted by the sectional shape in the midspan. This reduces the three-dimensional design problem into a two-dimensional one.

In [10], a genetic algorithm (GA) has been applied to optimise target pressure distributions around airfoils for inverse design methods. Pressure distributions are parameterised by B-spline polygons and the airfoil drag is minimised under constraints on lift, airfoil thickness and other design principles. Once target pressure distribution is obtained, corresponding airfoil geometry can be computed by an inverse design code by Takanashi [16] coupled with a Navier-Stokes solver [17].

Once the airfoil shape is designed, the next step of the wing design process is to determine the variation of the designed airfoil in the spanwise direction. The design principles for this step are essentially twofold. One is to preserve the two-dimensional performance as much as possible. This is easily achieved by the inverse method by specifying the same chordwise pressure distribution along the wing span. The other is to minimise the induced drag essential to the three-dimensional wing. The incompressible flow theory predicts that the minimum induced drag is achieved by an elliptical lift distribution (the lift per unit span varies elliptically along the span) [18]. Therefore, the elliptical lift distribution is the key design principle for three-dimensional wing shape optimisation.

The two design principles described above, however, contradict each other in general. Since the sectional lift is given by the chordwise pressure distribution, the elliptical lift distribution can be materialised by specifying the same chordwise pressure distribution along the wing span only if the wing has an elliptic planform. (The planform of a wing is defined as the shape of the wing when viewed from directly above.) Because of the manufacturing cost, however, modern commercial aircraft usually use a tapered wing instead of an elliptic wing.

Therefore, target pressure distributions should be optimised to minimise the induced drag, that is, to achieve the elliptical lift distribution for a tapered wing as well as to reduce the viscous drag for airfoil sections of the wing using the previous two-dimensional approach. This will lead to multiobjective optimisation. As a multi-objective GA (MOGA), we have adapted the Pareto-based ranking method by Fonseca and Fleming [19]. The design result will be given for a typical transonic wing.

9.1 Direct optimisation of airfoil

9.1.1 Approximation concept

The airfoil design is to determine the contour y for both upper and lower surfaces at every chordwise location. The way of expressing the y co-ordinates determines the choice of design variables. Following [1], let's store airfoil designs in vectors Y^1, Y^2,..., Y^N, where these vectors contain the co-ordinates of the upper surface followed by those of the lower surface at given chordwise locations. They correspond to the existing airfoil shapes as the basis vectors. Thus an airfoil shape is defined as:

$$Y = a_1 Y^1 + a_2 Y^2 + ... + a_N Y^N \qquad (9.1)$$

The design variables are now a_1 through a_N. This greatly reduces the number of design variables, say, compared with having the pointwise values of the y co-ordinates over 50 chordwise locations. In the following, four basic airfoil shapes are used ($N = 4$). As defined in [1], Y^1, Y^2, Y^3 and Y^4 indicate NACA2412, NACA64$_1$-412, NACA65$_2$-415 and NACA64$_2$A215, respectively.

9.1.2 Results of direct optimisation

To simplify the present aerodynamic optimisation problem, low speed airfoils are considered, assuming that the flow field is governed by the two-dimensional, incompressible, inviscid flow equation. A simple panel method described in [20] can be used for the flow analysis.

Now let's consider the lift maximisation problem. The objective function can be defined as the lift coefficient to be maximised. The only constraint used here is the maximum airfoil thickness, which must be 15 % of the chord. The angle of attack is fixed at six degrees for the flow analysis.

First, lift maximisation is considered with two design variables, a_1 for NACA2412 and a_2 for NACA64$_1$-412, for demonstration purposes (thus $a_3 = a_4 = 0$). Since only two design variables are used, we can easily visualise the distribution of the objective function, lift coefficient C_l, as shown in Figure 9.1 where $-5 < a_1$, $a_2 < 5$ and the lift coefficients are computed at an interval of 0.2 in both a_1 and a_2 coordinates. The negative value of the lift is replaced by zero.

The resulting distribution of the aerodynamic performance is highly irregular. Recall that both of the basis airfoils have 12 % thickness. The

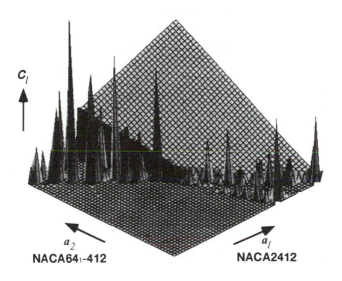

Figure 9.1   Lift distribution in design space

design variables are expected to approximately satisfy the relation:

$$a_1 + a_2 = \frac{15}{12} \tag{9.2}$$

to increase the airfoil thickness from 12 % to 15 %. However, the lift distribution is not smooth throughout eqn. 9.2 because of the different cambers and thickness distributions of the two basis airfoils. In addition, the Figure indicates that the maximum lift is achieved when $a_1 = a_2 = 5$. However, this is not acceptable because the airfoil thickness will be about 120 % of the chord. Since the flow equation is linearised and only two design variables are involved, one might expect a smooth distribution of the objective function. On the contrary, the resulting distribution has sharp, distinct, multiple peaks. This is a typical situation where GM will not work.

To see the effect of the constraint, the objective function is now redefined by using a penalty function with the airfoil thickness to chord t/c as:

$$F = C_l \cdot \exp[-100 \times |t/c - 0.15|] \qquad (9.3)$$

The corresponding function distribution is shown in Figure 9.2. Although the distinct, multiple peaks are greatly reduced, now most of the design space has zero objective function value. The design space with positive objective function values is found only in the narrow ridge along with eqn. 2 and it still contains five local extrema. This is another typical situation where GM will have a difficulty. In these situations, a mechanism to locate a global optimum is required. The combination of mutation and recombination of GA will be effective in finding an optimal solution. On the other hand, the hill-climbing strategy of GM is practically inapplicable to this problem.

Now the next test case considers lift maximisation with a full set of four design variables. The same constraint and flow condition were used. This time three optimisation methods, GM, SA and GA, were actually run for comparison. For GM, the feasible direction method in ADS V3.0, a FORTRAN program for automated design synthesis [21], was used. For SA, the code listed in [12] was used. For GA, the simple GA in [2] was adapted with the real number coding.

Figure 9.3 summarises the optimum lift values obtained from all three methods. For the GM and SA cases, four apparent initial designs are used. The lift coefficient of the GA case is 2·46, which is the best of the three and much higher than those of the previous two optimisation cases. Although both GA and SA are capable of getting out of local extrema, GA outperforms SA. This is because GA uses a population for simulated evolution, and SA uses a single design for annealing. Thus, the result of GA depends on the initial design less than that of SA. In addition, although SA produced consistent results against different initial points, the results of GM strongly depend on the initial points. GM failed to find an optimum when starting from the third initial design by reaching a negative thickness of the airfoil.

Figure 9.4 shows comparison of number of function evaluations required for the three methods. For the GM and SA cases, the sum of four runs (corresponding to four different initial designs) was plotted.

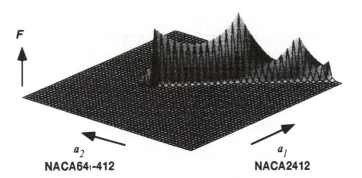

Figure 9.2   *Objective function distribution (lift coefficient with a penalty for airfoil thickness)*

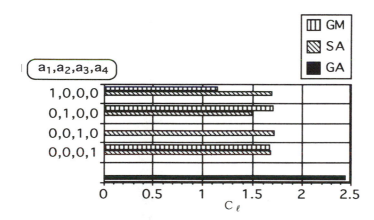

Figure 9.3   *Comparison of design results among GM, SA and GA*

It confirms that GA is the most time-consuming method. However, if we use GM or SA, we have to run it starting from many different initial designs to obtain a comparable result to that of GA. Since there are no guidelines for how to choose the initial design, it has to be an exhaustive, random search. As a result, GM and SA will not have any advantage in efficiency. Overall, GA is the best choice for this test case.

In general, aerodynamic performance is sensitive to geometries. So far, any theory of fluid dynamics cannot tell us how to choose design variables which will guarantee the convexness of the objective function. In compressible flows, the objective function itself may be discontinuous due to shock waves. Thus, the distribution of the objective function will be quite unexpected and the resulting aerodynamic optimisation problem will be very difficult. In this situation, a global search algorithm is indispensable and thus GA is preferred for the aerodynamic optimisation.

9.2 Inverse optimisation of the airfoil

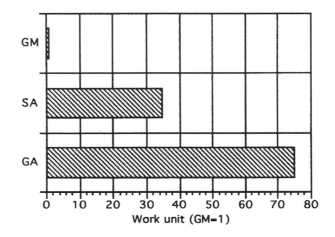

Figure 9.4   Comparison of number of function evaluations among GM, SA and GA

Target pressure optimisation for the airfoil is considered in this Section. As shown in the previous Section, GA will lead to the best solution for aerodynamic optimisation. However, it requires a tremendous number of computations. For engineering purposes, the direct approach will be too expensive. Therefore, inverse optimisation is developed to alleviate the computational time.

9.2.1 Coding

Genetic algorithms simulate evolution by selection. Design candidates are considered as individuals in the population. An individual is characterised by genes represented as a string of parameters. Here, an individual is a pressure distribution. Therefore, a coding scheme is required to specify a pressure distribution in terms of a string of parameters.

One of the parameterisation techniques recommended for *ab initio* designs is B-spline parameterisation [22]. The B-spline curve can be constructed so that the first and last points coincide with those of the defining polygon. Thus, pressure distributions are split into two curves, corresponding to pressure distributions on the upper and lower surfaces of an airfoil, respectively. As shown in Figure 9.5, seven points are used to define a B-spline polygon, specifying the pressure coefficient at the stagnation point $C_p = 1$ at $x = 0$ (since the inverse method does not solve the stagnation point exactly) and the pressure coefficient at the trailing edge $C_p = C_{p,te}$ (by user specification) at $x = 1$ (assuming the chord length unity).

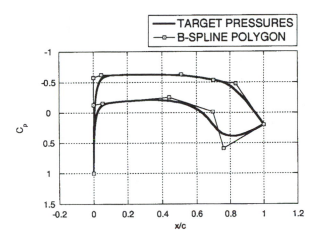

Figure 9.5   *B-spline polygons and corresponding pressure distributions*

Among 14 points for defining upper and lower C_p curves, ten points are free to move. The x and y ($y = C_p$) coordinates of those points are the design variables and thus we have 20 design variables in total. Although standard GAs are characterised by the use of the binary coding for design variables, a real (decimal) number is used for simplicity.

Initial population is generated randomly in the region of $0 < x < 1$ and $-1.5 < y < 1$. Besides the leading and trailing edges, one more point of the seven-point B-spline polygon is initially confined on the y axis to describe a steep C_p drop near the leading-edge region, that is, to obtain a large leading-edge radius typical for a supercritical airfoil [23]. To create each individual, the lower C_p curve is first generated based on constraints mentioned later. Then, the upper C_p curve is generated so as to satisfy approximately the rest of the constraints. If the resulting lift differs from the specified lift by less than 10 %, the corresponding string of parameters is assigned to an individual as genes. If it differs by more than 10 %, the string of parameters is discarded and regenerated. This process is repeated until 100 individuals are created.

9.2.2 Simple GA with real number coding

At each generation (iteration) of GA, the fitness value (objective function value) of every individual is evaluated and used to specify its probability of reproduction. A new population is generated from selected parents by performing specific operators on their genes. These operators are briefly explained in the following.

A simple GA is composed of three operators [2]: reproduction, crossover and mutation. Reproduction is a process in which individual strings are copied according to their fitness values. This implies that strings with a higher value have a higher probability of contributing one or more offspring in the next generation. A typical reproduction operator is the roulette wheel method described in [2]. The reproduction process produces a mating pool as a result. Then crossover proceeds in two steps. First, members in the mating pool are mated at random. Second, each pair of strings undergoes partial exchange of its strings at a random crossing site. This results in a pair of strings of a new generation. Mutation is a bit change of a string that will occur during the crossover process at a given mutation rate. Mutation implies a random walk through the string space and it plays a secondary role in the simple GA.

A simple crossover operator for real number strings is the average crossover [24] which would compute the arithmetic average of two real numbers provided by the mated pair. A weighted average can be used as:

$$\text{Child1} = r*\text{Parent1} + (1-r)*\text{Parent2}$$
$$\text{Child2} = (1-r)*\text{Parent1} + r*\text{Parent2} \qquad (9.4)$$

where Child1, 2 and Parent1, 2 denote 20 design variables of the children (members of the new population) and parents (a mated pair of the old generation), respectively. The uniform random number r ($0 < r < 1$) is regenerated for every design variable. Because of eqns. 9.4, a number of the initial population is assumed even.

Mutation takes place at a probability of ten per cent (when a random number satisfies $r1 < 0.1$). Eqns. 9.4 will then be replaced by:

Child1 = r*Parent1 + (1-r)*Parent2 + (r2-0·5)/5
Child2 = (1-r)*Parent1 + r*Parent2 + (r3-0·5)/5   (9.5)

where r2 and r3 are also random numbers for determining the amount of mutation.

9.2.3 Fitness evaluation: objective and constraints

In this Chapter, we define the optimisation problem as:

Minimise: drag coefficient C_d
Subject to: 1 lift coefficient C_l = specified
        2 airfoil thickness t = specified
        3 $C_{p,l} < 0$ at $0·1 < x < 0·6$ and $\int_{0.1}^{0.6} |C_{p,l}| dx \geq 0·1$
        4 $\max_{0.6<x<1} C_{p,l} < 0·4$
        5 $C_p|_{\text{suction peak}} \leq C_p^*$
        6 $\dfrac{dC_{p,u}}{dx}\bigg|_{0.1<x<0.5} \cong 0$
        7 $\dfrac{dC_{p,u}}{dx} \leq 2·5$
        8 number of inflection points < 2

To specify airfoil thickness t approximately, a formula is taken from [15] as:

$$t = \frac{-\sqrt{1 - M_\infty^2}}{2} \int_0^2 \frac{C_{p,u} + C_{p,l}}{2} dx \tag{9.6}$$

where M_∞ is the freestream Mach number. Constraints 5 and 6 aim to reproduce the sonic-plateau pressure distributions described in [23]. Constraint 7 was taken from the observation of C_p plots in [23] to avoid a separation of the boundary layer.

Drag is calculated as the sum of viscous and wave drag. When the flow is attached, the profile drag can be calculated from a knowledge of the potential flow pressure distributions and location of transition from laminar to turbulent flow. Locations of transition are left for user specification. When no shock wave is present, wave drag is ignored.

The Squire-Young relation is an empirical relation between drag based on the momentum thickness of the boundary layer and the potential velocity [25]. The momentum thickness can be estimated from an integral equation of the turbulent boundary layer. The viscous drag is estimated as

$$C_{dv} = 2\theta_{te}\left(\frac{U_{e,te}}{U_\infty}\right)^{3.2} \qquad (9.7)$$

where θ_{te} is the momentum thickness at the trailing edge, $U_{e,te}$ is the potential velocity at the trailing edge and U_∞ is the freestream velocity. See [25] for more detail.

The optimisation took about three minutes on an SGI Indy workstation using 1000 successive generations with 100 individuals in the population. It is far less than the computational time necessary for the inverse design as mentioned later.

9.2.4 Construction of fitness function

GAs need to define an objective to be maximised. An inverse of the drag coefficient is taken as the objective here. Then, all the constraints are required to be combined with the objective. There are eight constraints in the fitness evaluation Section. Constraints 1 and 2 are equality constraints and constraints 3 to 8 are inequality constraints. The equality constraints are multiplied to the objective as an exponential function to reject the individuals which do not satisfy the specified lift and airfoil thickness. The inequality constraints are expressed so as to increase their values when violated and the inverse of their sum is added to the objective with penalty.

The final form of the objective is given as:

$$Fitness = \left(\frac{0.4}{C_d^2} + \frac{2 \times 10^7}{IC^2}\right) \cdot \exp(-100 \times EC) \qquad (9.8)$$

where IC and EC denote the sum of inequality constraints and equality constraints, respectively. For IC, constraints 3 to 8 are represented as:

$$IC = 10000 \times \left[\min\left(\int_{0\cdot1}^{0\cdot6} |C_p| \, dx, \, 0\cdot1 \right) - 0\cdot1 \right]^2$$

$$+5 \times \left[\max_{0\cdot6<x<1}(C_{p,l}, \, 0\cdot4) - 0\cdot4 \right]^2$$

$$+3 \times \left[\max\left(\frac{C_p|_{\text{suction peak}}, \, 1\cdot0}{C_p^*} \right) - 1\cdot0 \right]^2$$

$$+40000 \times \left[1.1 - \min\left(\frac{slope + |slope|}{2}, 1\cdot1 \right) \right]$$

$$+0\cdot001 \times \left[\max\left(\frac{dC_{p,u}}{dx}, 2\cdot5 \right) - 2\cdot5 \right]^2$$

$$+0\cdot01 \times \exp[\text{Number of inflection points}]$$

(9.9)

where $slope = \dfrac{C_{p,u}|_{x=0\cdot5} - C_{p,u}|_{x=0\cdot1}}{0\cdot4} + 1$. For *EC*, constraints 1 and 2 are represented as:

$$EC = \max\left(|t_{\text{specified}} - t_{\text{calculated}}|, \, 10^{-4} \right)$$

$$+ \max\left(|C_{l\,\text{specified}} - C_{l\,\text{calculated}}|, \, 10^{-4} \right)$$

$$-0\cdot01$$

(9.10)

where the differences below 10^{-4} are ignored for the optimisation.

9.2.5 Inverse design cycle

Once the present GA finds an optimum target pressure distribution, a corresponding airfoil geometry can be obtained by an inverse design method (Figure 9.6). Here the inverse design code WinDes is used [16]. The code can solve both two- and three-dimensional problems.

WinDes uses the following iterative procedure. Suppose the initial geometry and surface pressure distributions obtained from any CFD analysis code are given. First, pressure differences are calculated from the given initial and target pressure distributions. From these pressure differences, corresponding geometry corrections can be computed from the integral equations discretised at the panels on the geometry. An improved geometry is then obtained from the initial geometry and the computed geometry corrections. Finally, the CFD code is used again to check how close the resulting pressure distributions are to the target distributions. If the differences are still large, the process will be iterated. In practice, ten to 20 iterations are sufficient to obtain the final geometry.

The advantage of this method is that the required analysis code is arbitrary and any type of analysis method, even experimental, can be used. In this Chapter, two and three-dimensional Navier-Stokes codes, LANS2D [26] and LANS3D [17], were used. The latest version of these codes uses the third-order upwind in the right-hand side.

In the present inverse design method, grid generation around the

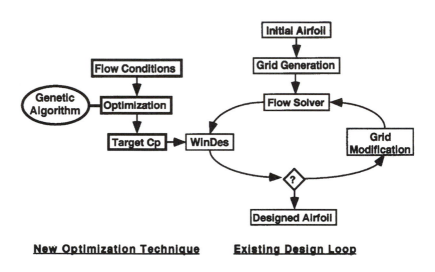

Figure 9.6   Flowchart of the present design procedure

modified geometry is required at every iteration. In order to automate the inverse design loop, a grid generator has to be robust and efficient. An algebraic grid generation code described in [26] is used because of its robustness and efficiency. The two-dimensional C-type mesh contains 131 times 51 grid points in the chordwise and normal (to the surface) directions, respectively. In the three-dimensional case, the C-H topology is used, applying the two-dimensional grid generation at each spanwise section. The three-dimensional grid contains 30 sections in the spanwise direction.

Use of WinDes, LANS2D/3D, and the algebraic grid generator constructs an automated loop for the inverse design with reasonable computational requirements. These codes were implemented on a

CRAY C90/161024 supercomputer at Institute of Fluid Science, Tohoku University. The inverse design of an airfoil for ten cycles required about 30 minutes on C90 using a single processor.

9.2.6 Results of airfoil design

In the first test case, the flow condition was set to the freestream Mach number of 0·75 and the Reynolds number of ten million. In the Navier-Stokes computation, the Baldwin-Lomax turbulence model [27] was used. Angle of attack was set to zero. Locations of transition were fixed at five and ten per cent chord for upper and lower surfaces, respectively. The lift was specified as 0·5 and the trailing-edge pressure coefficient was set to 0·15. Figure 9.7 shows the optimisation history of the present GA. The optimum was obtained after about 450 generations.

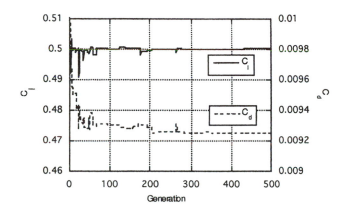

Figure 9.7   *Optimisation history of lift and drag of the best fit in generation*

Figure 9.8 shows the target pressure distribution obtained from the present GA, the designed geometry obtained from the inverse method and the corresponding pressure distribution obtained from the Navier-Stokes computation. The optimised pressure distribution has a sonic plateau to avoid a shock wave on the upper surface of the airfoil and a rear loading region typical for the supercritical airfoils. In the inverse design cycle, the initial geometry was chosen as the NACA0012 airfoil. In total nine iterations were required to obtain the final geometry.

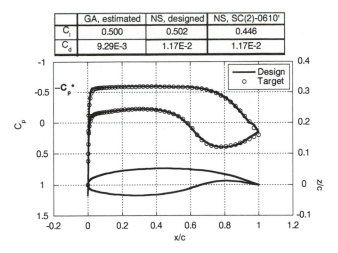

	GA, estimated	NS, designed	NS, SC(2)-0610'
C_l	0.500	0.502	0.446
C_d	9.29E-3	1.17E-2	1.17E-2

Figure 9.8   *Design results and comparison of aerodynamic performance for the airfoil case*

In Figure 9.8, the attached table shows a comparison of the aerodynamic performances among the target (indicated as GA, estimated), the inverse design result (indicated as NS, designed) and a modified supercritical airfoil (indicated as NS, SC(2)-0610'). The SC(2)-0610 airfoil is taken from [23]. Although it has a blunt trailing edge, the geometry is modified to have a sharp trailing edge for comparison purposes. The drag obtained from the target optimisation was underestimated due to the simplified calculation of the viscous drag. Comparing the NS (Navier-Stokes) results, the present design shows higher L/D (lift to drag) ratio than the modified SC(2)-0610 airfoil.

9.3 Inverse optimisation of the wing

Once the airfoil shape is designed, the next step of the wing design is to determine the variation of the designed airfoil in the spanwise direction. The design principles for this step are essentially twofold as mentioned at the beginning of this Chapter. One is to preserve the two-dimensional performance as much as possible. This is easily achieved by the inverse method by specifying the same chordwise pressure distribution along the wing span. The resulting wing has the straight isobar pattern of pressure contours on the wing surface.

The other is to minimise the induced drag. If the wing has lift, the average pressure over the bottom surface of the wing is greater than that over the top surface. Consequently, there is some tendency for the air to flow around the wingtips from the high to low-pressure sides. This flow establishes wingtip vortices. These vortices induce a small downward component of air velocity in the neighbourhood of the wing itself. Because the local relative wind is canted downward, the lift vector itself is tilted back; hence it contributes a certain component of force parallel to the freestream, that is, a drag force.

Since the induced drag becomes one half to two thirds of the total drag during climb, reduction of the induced drag is an important goal for the three-dimensional wing design. According to the incompressible flow theory, the minimum induced drag is achieved by an elliptical lift distribution [18]. Therefore, the elliptical lift distribution is the key design principle for wing shape optimisation.

The induced drag is strongly influenced by the planform of a wing. Planform is directly related to aspect ratio and taper ratio of a wing. Taper ratio has a great effect on the spanwise lift distribution and thus there is an optimal taper ratio for a wing to minimise the induced drag [28]. Tapered wings have been adopted for the majority of aircraft nowadays since they offer a compromise solution on account of their low induced drag, high maximum lift, low structural weight, good stowing provisions for the undercarriage and reasonable manufacturing cost. The present design method will minimise the induced drag for any taper ratio and thus it will provide more design opportunities for wing shapes.

9.3.1 Pressure distribution for the wing

Target pressure distribution for the three-dimensional wing can be obtained by specifying the chordwise pressure distributions at several spanwise sections. Planform shape of a wing is usually determined by other means and thus a typical wing planform of a transonic transport aircraft is assumed here.

The present objective of the wing design is to minimise the induced drag. This is achieved by elliptical lift distribution in the spanwise direction of the wing. The constraint in the total lift will specify an elliptical lift distribution uniquely. Thus, the objective function can be given by differences of the sectional lifts to the elliptic distribution at the several spanwise sections. The three-dimensional optimisation problem is now defined as:

Minimise: 1 difference of the spanwise lift distribution to the
              elliptic distribution

                2   two-dimensional drag coefficient C_d at each spanwise
                    section
Subject to: additional constraints for chordwise pressure distribu-
                tion at each spanwise section

We can further redefine the constrained problem to the unconstrained multiobjective optimisation problem as:

Minimise: 1   difference of the spanwise lift distribution to the
                elliptic distribution
            2   two-dimensional drag coefficient C_d at each spanwise
                section
            3   penalty function for chordwise pressure distribution at
                each spanwise section in Section 9.2.4.

9.3.2 MOGA

Before implementing the Pareto ranking approach for the present
MOGA, we have tried a few other ways of constructing a GA for the
present multiobjective optimisation. First, a simple GA was used by
combining three objective functions into a single one. However, this
approach not only failed to search Pareto-optimal solutions, but also
produced premature convergence. Certain spanwise sections had
unacceptable chordwise pressure distributions for the airfoil section.
Next, the vector evaluated genetic algorithm (VEGA) [29] was
adapted to the present problem. As pointed out in [30], however, the
solution was extremely good for one objective but not for the others.
These experiences led us to Fonseca-Fleming's Pareto ranking method
[19].

   In the present MOGA, the third objective for the penalty function is
used to pool the top 30 % individuals in the population. Then
Fonseca-Fleming's Pareto ranking method is applied to these
individuals by using the first and second objectives. A selection
operator is defined by using the nonlinear function suggested in [31].
Crossover and mutation operators are defined similar to those in
Section 9.2.2. The elite strategy is also used to preserve the best
individual for each objective. After 200 generations, the best solution
in terms of the first objective is selected from the Pareto-optimal set as
the optimal solution.

   As mentioned in Section 9.2.1, random creation of initial
population produces infeasible solutions due to the severe constraints.
Thus, we first ran the two-dimensional GA by using only the

constraints to evolve a population of feasible solutions. Then we distributed the sectional pressure distribution to the six spanwise sections from the root to the 83·3 % span so as to give the elliptical lift distribution approximately. To do this, we only changed the pressure on the lower surface of the airfoil. In this way, we were able to implicitly satisfy the first design principle for the wing mentioned in the beginning, that is, to maintain the two-dimensional performance. The straight isobar pattern of pressures on the upper surface of the wing is expected to produce drag divergence at the same Mach number along the wing span and thus the resulting drag divergence Mach number of the wing will be similar to that of the airfoil section. The population of 210 individuals was used as the initial population of the present MOGA.

Once the present MOGA finds an optimum target pressure distribution, corresponding wing geometry can be obtained by an inverse design method similar to the airfoil case. The inverse design code, Navier-Stokes code, and algebraic grid generator were implemented on an NEC SX-4 supercomputer at the Department of Aeronautics and Space Engineering, Tohoku University. The inverse design for one cycle required about 45 minutes of single CPU time (most of the time is used for the Navier-Stokes computation).

9.3.4 Results of wing design

As a model wing for a transonic transport aircraft, the simple, swept and tapered wing shown in the left-hand side of Figure 9.9 is considered for shape optimisation. The wing has a sweep angle of 20·4 deg, an aspect ratio of 7·38 and a taper ratio of 0·3. It should be noted that this taper ratio is small because such a wing has approximately an elliptical lift distribution.

The elliptical lift distribution was monitored at six locations from the root to the 83·3 % span, as indicated. The inverse solver used the same spanwise locations for the geometry correction. For the Navier-Stokes grid, the modification of wing geometry was linearly interpolated between those sections. In the tip region, the same airfoil section was used outside of the 83·3 % section, while the wing twist was linearly extrapolated. The tip region is usually designed by other means and thus the optimisation of this region is not considered here.

The right-hand side of Figure 9.9 shows the computed pressure contours on the upper surface of the wing designed by the inverse method based on the target pressure distribution optimised by the present MOGA. The flow condition had a freestream Mach number of

0·75, the Reynolds number based on the root chord of 10⁷ and an angle of attack of 0 deg. The resulting straight isobar pattern satisfies the first design principle well and thus indicates good performance at higher Mach numbers. On the other hand, it shows a minor oscillation near the leading edge toward the root section. Although the airfoil sections vary very much from the root to 16·7 % section, a linear interpolation is used to create a Navier-Stokes grid for brevity. To treat the root region as well as the tip region more precisely, an elaborate procedure may be necessary.

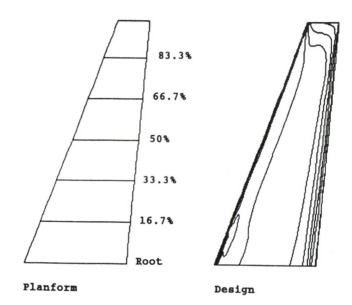

Figure 9.9   *Wing planform and computed pressure distribution on the designed wing*

Figure 9.10 shows the computed lift distribution of the designed wing in comparison to the elliptic distribution. The result is found to satisfy the second design principle closely. Figure 9.11 shows the target chordwise pressures obtained from the present MOGA, the resulting airfoil shape of the wing and the corresponding pressures computed by the Navier-Stokes solver at the 16·7 %, 50·0 % and 83·3 % spanwise sections. It confirms that the inverse problem is solved satisfactorily except at the leading edge near the root section. The discrepancy of the pressure profiles there corresponds to the oscillation found in Figure 9.9.

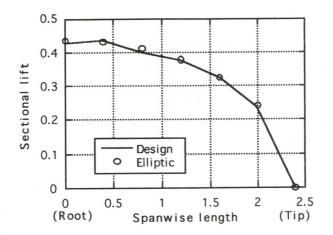

Figure 9.10   Sectional lift distribution in the spanwise direction

Figure 9.12 summarises the aerodynamic performance of the designed wing in terms of the lift to drag ratio. For comparison purposes, two other wings were designed by the inverse method. The design indicated as 'alternate' was obtained by changing the upper surface pressures when distributing the two-dimensional pressure distributions to the six spanwise sections for the initial population. Then the same MOGA was run. This procedure allows a wide variation in the pressure distributions in the spanwise direction. The resulting wing satisfies the second design principle, for an elliptic distribution, better than the present design but not the first principle. Thus, it performs better at the design point but worse at higher Mach numbers.

The other design indicated as 'isobar' was obtained by specifying a straight isobar pattern on both upper and lower surfaces of the wing. The resulting wing satisfies the first design principle of the wing exactly but not the second one. However, this is the standard design procedure for transonic wings. The reduction of the induced drag simply relies on the use of a proper taper ratio. In fact, due to the present taper ratio of 0·3, this wing gives good performance similar to the present design. Since the geometries of the two are completely different, this result confirms that the present design gives a Pareto-optimal solution under the contradicting design principles for the wing.

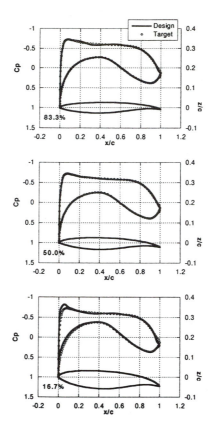

Figure 9.11   Designed airfoil sections and corresponding pressure distributions

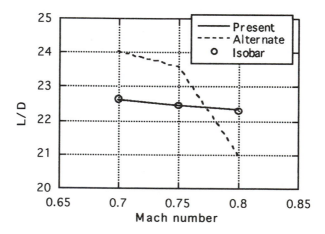

Figure 9.12   Comparison of L/D performances

9.4 Summary

Characteristics of aerodynamic optimisation have been discussed through wing shape design problems. It has been demonstrated that distribution of the objective function can be extremely rough even in a simplified problem. In such a situation, GA (genetic algorithm) is expected to be more effective than a simple hill-climbing strategy.

Three optimisation algorithms, the GM (gradient-based method), SA (simulated annealing) and GA, were first applied to the airfoil shape design using the approximation concept to compare their performances. Although GA is time consuming, its result is superior to those of the others. Since the other algorithms will require many trials starting from various initial designs to obtain a comparable result, they will not have any advantage in efficiency. The result suggests that GA is the best option for aerodynamic optimisation.

To alleviate the large computational time necessary for GA, the inverse optimisation method has been developed to optimise target pressure distributions. GA is applied to find a pressure distribution which minimises the airfoil drag under constraints on lift, airfoil thickness and other design principles. Once the target pressure is given, the corresponding geometry can be found by an inverse code coupled with a Navier-Stokes solver.

MOGA based on Fonseca-Fleming's Pareto ranking method has been developed to optimise the three-dimensional target pressures for the aerodynamic inverse design of wing shape. The optimisation problem was formulated to minimise the induced drag for the wing as well as to minimise the viscous drag for airfoil sections. Performances of both the simple GA and VEGA were found unsatisfactory to the three-dimensional optimisation problem.

The resulting procedure was successfully applied to transonic wing design. The standard design procedure for transonic wings has previously focused on materialising the straight isobar pattern over the wing. Reduction of the induced drag merely relied on the use of a proper taper ratio for the wing planform.

The present design procedure allows the minimisation of the induced drag for an arbitrary wing planform with any taper ratio. This will provide more design opportunities for wing shapes in terms of better aerodynamic performance, lighter structural weight and less expensive manufacturing costs.

9.5 References

1   Vanderplaats, G. N.,: *Numerical otimization techniques for engineering design: with applications* (McGraw-Hill, Inc., New York, 1984)

2   Goldberg, D. E.: *Genetic algorithms in search, optimization & machine learning* (Addison-Wesley Publishing Company, Inc., Reading, Jan. 1989)

3   Bramlette, M. F., and Cusic, R.: 'A comparative evaluation of search methods applied to the parametric design of aircraft'. Proceedings of the third international conference on *Genetic alrgorithms* (Morgan Kaufmann Publishers, Inc., San Mateo, June 1989) pp. 213–218

4   Crispin, Y.: 'Aircraft conceptual optimization using simulated evolution'. AIAA paper 94-0092, Jan. 1994

5   Powell, D. J., Tong, S. S,, and Skolnick, M. M.: 'EnGENEous domain inependent, machine learning for design optimization'. Proceedings of the third international conference on *Genetic alrgorithms* (Morgan Kaufmann Publishers, Inc., San Mateo, June 1989) pp. 151-159

6   Gage, P., and Kroo, I.: 'A role for genetic algorithms in a preliminary design environment'. AIAA paper 93-3933, August 1993

7   Gregg, R. D., and Misegades, K. P.: 'Transonic wing optimization using evolution theory'. AIAA paper 87-0520, Jan. 1987

8   Quagliarella, D., and Cioppa, A. D.: 'Genetic algorithms applied to the aerodynamic design of transonic airfoils'. AIAA paper 94-1896, June 1994

9   Yamamoto, K., and Inoue, O.: 'Applications of genetic algorithm to aerodynamic shape optimization'. AIAA paper 85-1650-CP, a collection of technical papers, 12th AIAA *Computational fluid dynamics* conference, CP956, San Diego, CA, June 1995, pp. 43-51

10   Obayashi, S; and Takanashi, S.: 'Genetic optimization of target pressure distributions for inerse design methods,' *AIAA J.* **34**, (5) pp. 881–886, 1996

11   van den Dam, R. F., van Egmond, J. A., and Slooff, J. W.: 'Optimization of target pressure distributions'. Special course on inverse methods for airfoil design for aeronautical and turbomachinery applications, AGARD Report 780, reference 3, Nov. 1990

12   Press, W. H., *et al.*: *Numerical recipes in FORTRAN: the art of scientific computing* (Cambridge University Press, Cambridge, 1992, 2nd edn.)

13   Labrujére, Th. E., and Slooff, J. W.: 'Computational methods for the aerodynamic design of aircraft components', *Ann. Rev. Fluid Mech.* **25**, pp.183–214, l993

14   van den Dam, R. F.: 'Constrained spanload optimization for mimum

drag of multi-lifting surface configuration'. Computational methods for aerodynamic design (inverse) and optimization, AGARD conference proceedings 463, reference 16, March 1990

15 van Egmond, J. A.: 'Numerical optimization of target pressure distributions for subsonic and transonic airfoil design'. Computational methods for aerodynamic design (inverse) and optimization, AGARD conference proceedings 463, reference 17, March 1990

16 Takanashi, S.: 'Iterative three-dimensional transonic wing design using integral equations', *J. Aircr.* **22**, (8) pp. 655–660, 1985

17 Fujii, K., and Obayashi, S.: 'Navier-Stokes simulations of transonic flows over a practical wing configuration', *AIAA J.*, **25**, (3), pp. 369–370, 1987

18 Anderson, Jr., J. D.: *Introduction to flight* (McGraw-Hill Inc., NY, 1989) pp. 216–222

19 Fonseca C. M., and Fleming, P. J.: 'Genetic algorithms for multiobjective optimization: formulation, discussion and generalization'. Proceedings of the 5th international conference on *Genetic algorithms* (Morgan Kaufmann Publishers, Inc., San Mateo, July 1993), pp. 416–423

20 Katz, J., and Plotkin, A.: *Low speed aerodynamics: from wing theory to panel methods* (McGraw-Hill Inc., New York, 1991, international edn.)

21 Vanderplaats, G. N.: 'ADS – a FORTRAN program for automated design synthesis, version 3.00'. 'Engineering Design Optimization, Inc., 1988

22 Rogers, D. F., and Adams, J. A.: *Mathematical elements for computer graphics* (McGraw-Hill, Inc., New York, 1990, 2nd. edn.)

23 Harris, C. D.: 'NASA supercritical airfoils – a matrix of family-related airfoils'. NASA TP-2969, March 1990

24 Davis, L.: *Handbook of genetic algorithms* (Van Nostrand Reinhold, 1990)

25 Young, A. D.: *Boundary layers* (AIAA Education Series, Washington, D. C., 1989)

26 Matsushima, K., Obayashi, S., and Fujii, K.: 'Navier-Stokes computations of transonic flow using the LU-ADI method'. AIAA paper 87-0421, Jan. 1987

27 Baldwin, B. S., and Lomax, H.: 'Thin-layer approximation and algebraic model for separated turbulent flows'. AIAA paper 78-257, January 1978

28 Torenbeek, E.: *Synthesis of subsonic airplane design* (Kluwer Academic Publishers, Dordrecht, 1982) pp. 232–237

29 Schaffer, J. D.: 'Multiple objective optimization with vector evaluated genetic algorithm'. Proceedings of the 1st international conference on *Genetic algorithms*, 1985, pp. 93–100

30 Tamaki, H., Kita H., and Kobayashi, S.: 'Multi-objective optimization by genetic algorithms: a review'. Proceedings of 1996 IEEE international conference on *Evolutionary computation*, 1996, pp. 517–522

31 Michalewicz, Z.: *Genetic algorithms + data structures =evolution programs* (Springer-Verlag, Berlin, 1994, 2nd extended edn.) pp. 57–58

Chapter 10
Genetic design of VLSI layouts
V. Schnecke

10.1 Introduction

Genetic algorithms (GAs) are well known as a robust optimisation method for a large range of design applications. This robustness is caused by the fact that a genetic algorithm works with a coding of the optimisation problem rather than the problem itself. It deals with a set of individuals, which represent candidate solutions to the optimisation problem. Generally, there is a distinction between the phenotype representation, which defines the real appearance of the individual, and the genotype, which encodes the genetic information needed for a full characterisation of the solution.

In many applications a set of continuous parameters has to be optimised. There the genotype representation usually is a string of genes (bits or floats) that define the values for those parameters. The genetic algorithm traverses the space of the representations by creating new individuals out of others when combining parts of the strings during recombination, or mutating single genes randomly. The individuals created by those operations are correct codings of admissible solutions and build the population for the next generation.

For some combinatorial optimisation problems a string-type genotype coding is also possible, but not every string represents a feasible solution. When the gene strings (chromosomes) of two individuals are crossed during recombination, it is unlikely that the resulting offspring represent correct solution encodings. There are generally two ways of dealing with this problem: incorrect individuals might be accepted and inserted into the population. When computing their fitness, a penalty must be added to take care that these individuals get a lower chance during the selection process. This yields convergence to a population of only correct individuals during the optimisation. Another way of getting rid of the incorrect individuals is to prevent their formation by including problem-specific knowledge in the operators such that they always produce correct offspring. This is in contrast to the definition of the simple GA introduced in the common textbooks [8, 11]. It is illusory to see the GAs as an optimisation tool which can solve any problem with standard operators, after encoding its solutions in a string of 0s and 1s. For real-world discrete optimisation problems much more effort has to be taken, as will be shown in this Chapter.

The application that will be described represents a combinatorial optimisation problem in the design cycle for VLSI-chips. Modules of a chip have to be placed on the layout surface and signal nets have to be routed on the space between these modules. The phenotype is a complete layout which defines the geometrical arrangement of the modules and the routes for the interconnection wires. Such a complex phenotype cannot be encoded in a string of elementary datatypes. The chosen genotype representation is a tree with additional information for all nodes defining details for routing and sizing of the modules. Although the execution of the genetic operators on a tree structure is rather straightforward, care must be taken to produce only correct offspring. In addition to this, there are some examples of exploiting problem-specific knowledge during the application of the operators. Together with a hybrid method for the creation of the initial individuals, the described approach contains many detailed features to make it work. But in most cases this is the only way to enable genetic algorithms to deal with real-world combinatorial optimisation problems.

10.2 Physical VLSI design

VLSI (very large scale integration) refers to a technology which enables the integration of more than a million transistors on a single chip. The design cycle for these VLSI chips consists of different consecutive steps from high-level synthesis (functional design) to production (packaging) [19]. All these steps contain many non-numerical, compute bound problems. Conceivably the most complex task of all is the physical design. This process can be characterised as the transformation of a circuit description into a physical layout. The layout describes the geometric representation of all components of the circuit and the shapes of the interconnection wires. Out of this layout, the masks for the different layers needed for fabrication are constructed.

Treating circuits with some hundreds of thousands of transistors has only become possible by the availability of CAD-tools. Because of the growing complexity of the design process, there is a strong interest in tools for physical design automation with no or only little human interaction. Automation of the design process increases the level of integration and enhances the chip performance. Further, it reduces the time to market, which is perhaps the most important reason for the steady strong interest in new approaches to VLSI physical design automation.

10.2.1 Macro cell layouts

For the layout design of microprocessors, which represent the largest circuits, the macro cell design style is used. This is the most complex style, and is mainly applied to mass produced chips, because it yields highly optimised layouts. Semicustom layout generation, used for standard cells or gate arrays, deals with more restrictions, thus reducing complexity of the design task, but also decreases the quality (level of integration) of the layouts [13].

At the beginning of the physical design process the circuit is partitioned to generate some ten macro cells, which represent functional units. On the border of each cell, terminals (pins) are

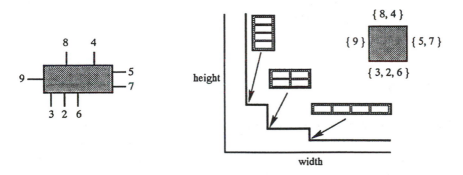

Figure 10.1   A fixed cell (left) and a flexible cell with the corresponding shape function (right)

located for the connection of the signal nets. A netlist specifies the connectivity of the cells. From the designers' point of view there exist two different kinds of macro cells: fixed cells have fixed dimensions and fixed terminal positions on their borders (Figure 10.1, left); flexible cells have different implementations with different aspect ratios, resulting from the hierarchical construction of the cells. These implementations are defined either by the area and upper and lower bounds for the width or height of the cell, or by a shape function [12], which describes the admissible shapes, i.e. the relation of area and aspect ratio. Figure 10.1, right, shows a flexible cell which consists of four subcells. The shape function is characterised by three minimal shapes, but every point above the curve defines an admissible implementation for this cell. Because the terminal positions of a flexible cell vary with its different implementations, in this case only a list of nets for each of its borders is given.

During layout generation the cells have to be placed on the layout surface and the signal nets must be routed on the space between the cells. At the border of the layout, pads have to be placed for the I/O connections of the chip. The main objective is the minimisation of the area of the rectangle which circumscribes all components; further there are timing constraints that enforce maximal admissible wirelengths for some nets. Due to its complexity, the generation of macro cell layouts is usually done in various substeps:

1   During the floorplanning phase the cells have to be placed on the layout surface and exact implementations for the flexible cells must be chosen.
2   After placement, the global routing is done. In this step the loose routes for the signal nets are determined.
3   In the detailed routing, the exact routes for the interconnection wires in each channel between two macro cells must be computed.
4   The last step is the compaction of the layout, where it is compressed in all dimensions so that the total amount of area is reduced.

This classical approach to layout generation is strongly serial with many interdependencies between the substeps. For example, while placing the cells, there must be an estimated amount of space reserved around them to enable the completion of the routing later on. In the case of fault estimation, the routing cannot be completed inside the reserved routing regions. This is either realised during the computation of the global routes, or during channel routing. In the latter case the layout generation process has to backtrack and different global routes for some nets have to be chosen. If even global routing is impossible, the cells need to be rearranged, i.e. the process has to backtrack to the floorplanning task.

As mentioned before, the division of the layout generation process results from the complexity of the whole optimisation problem. Nevertheless, most of the sub-tasks are still intractable so that only heuristic approaches can be used.

10.2.2 Placement

During placement the macro cells are positioned on the layout surface in such a manner that no cells overlap each other and that there is enough space left between them to complete the interconnections later on. Floorplanning is a generalisation of the placement problem. It deals with flexible cells which have to be sized during (or after) placement to yield a layout with an overall minimal area. A floorplan is a partitioning of the layout surface into different rooms, inside which the flexible cells are placed.

Several approaches to the placement or floorplanning problem exist. Here an overview of the (nongenetic) solution methods is given; for more details see the textbook of Sait and Youssef [13] or the survey of Shahookar and Mazumder [18].

One of the early techniques is force-directed placement. Here, cells that are connected by common nets exert an attractive force which is proportional to the number of these nets and the distance between the cells. The ideal positions for the cells are computed numerically by solving a set of equations, which corresponds to finding an equilibrium in a minimum energy state.

Partition based methods recursively divide the set of cells. At the same time, the available chip area is partitioned and each set of cells is assigned to one of the components. By using a mincut heuristic, the number of cut nets in each division is minimised, which yields highly connected cells to be placed near together. The major drawback of this approach is that the locations of the external connections are not considered during placement of a subcircuit.

A very popular technique for computing the placement is simulated annealing [17, 20]. This is an iterative improvement method, which simulates the behaviour of atoms during the cooling schedule of molten metals. It starts with a randomly created arrangement of the cells and successively computes new configurations by moving or exchanging cells. A new solution is accepted depending on its quality and on the temperature. During the early stage of optimisation, when this value is rather high, even inferior new solutions are accepted, which prevents the optimisation from getting stuck in a local optimum. Simulated annealing yields high quality placements but with an excessive amount of computation time.

10.2.3 Routing

The aim of the routing phase is to find the geometrical layouts for all nets. During the floorplanning phase, space on the layout surface is provided for the routing of the signal nets. This space can be described as a collection of routing regions. Each region has a fixed capacity, i.e. a maximum number of wires that can be routed through this region, and a number of terminals, i.e. pins on the borders of the adjacent cells.

Due to the complexity, the routing is done in two subphases. In global routing, each net is assigned to particular routing regions. A common way to compute this allocation is based on graph algorithms. Here the total routing space is described by a graph: the edges of this graph represent the routing regions and are weighted with the corresponding capacities; the vertices are the connections between two regions. Global routing is described by a list of routing regions for each net of the circuit, with none of the capacities of any routing region being exceeded (Figure 10.2). The main objective is to find a global routing with a minimum estimated overall wiring length. For two terminal nets shortest paths algorithms are used, and for nets with three or more terminals a minimum rectangular steiner tree is computed. The described technique is usually cited as the sequential approach to the global routing; other approaches are based on integer-programming, hierarchical decomposition, or random search techniques such as simulated annealing or genetic algorithms [10, 13].

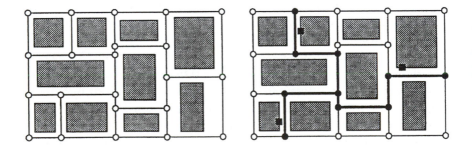

Figure 10.2   A routing graph (left) and a global route for a three terminal net (right)

After global routing; is completed, the number of nets routed through each routing region is known. In the detailed routing phase, the exact shapes for the wires have to be determined (Figure 10.3). This is done incrementally, i.e. one channel is routed at a time in a predefined order.

The complexity and routability of a layout depends on the number of layers which can be used for the completion of the inter-

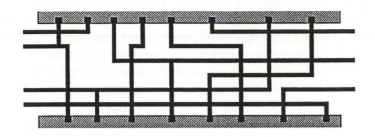

Figure 10.3   The detailed routing inside a channel

connections. In macro cell layouts there are usually two layers and the routing is done using the manhattan-model, i.e. there are only horizontal and vertical line segments. The simplest model is the restricted manhattan-routing, where one layer is used for the vertical, the other for the horizontal wires and the nets change the layer when changing their direction. For detailed routing there exist solutions based on greedy methods, graph algorithms or hierarchical approaches [13, 19].

10.2.4 Previous genetic approaches

Although VLSI design is usually cited as a typical application domain for GAs, there only exists a handful of research papers dealing with layout optimisation.

The classical work has been done by Cohoon *et al.* [3, 4]. They present a parallel genetic algorithm for floorplan design and use a weighted sum of the total area and the estimated wirelength as the objective function. A placement is described by a binary slicing tree with the leaves representing the cells and the inner nodes defining the cut directions. The genotype is encoded as a normalised polish expression resulting from a post-order traversal of this tree. They have implemented different recombination operators, which work either on this string without considering the structure of the tree, or directly transmit subtrees from the parents to the offspring. Results are presented for artificial circuits with up to 25 cells.

Chan *et al.* [2] introduced a bit-matrix representation. The layout area is divided into a number of quadratic regions. Placement for a single cell is represented by binary encoding of the information about the occupied squares and the orientation of this cell. The total placement information is combined in a bit matrix, each line in this matrix describing the placement for a single cell. Cells are allowed to overlap each other during optimisation, which is approached by adding a penalty to the fitness of an incorrect individual. Routing is included by estimating the wirelength and crossover is done by constructing an offspring out of the quarterised parent matrices. The authors present results for benchmark circuits with up to 49 cells.

Esbensen [5] describes a GA for macro cell placement where the genotype is encoded as a binary tree, but his approach is not restricted to slicing floorplans (Figure 10.4). Each node of the tree represents a cell and, due to a given node order, a placement can gradually be generated by decoding the genotype. The quality of a placement is determined by the layout area; for layouts with equal area an estimated wirelength is taken into consideration. Later Esbensen and Mazumder [6] combined this algorithm with simulated annealing which produced better results than the previous GA. Due to complexity constraints, only (real) circuits with up to 11 cells and 203 nets are computed.

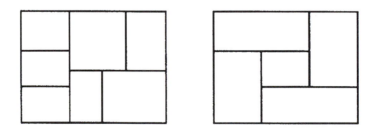

Figure 10.4   A slicing (left) and a nonslicing floorplan (right)

10.3 A GA for combined placement and routing

As described in Section 10.2.1, the global routing is usually computed after placing the modules, and during placement an estimated amount of routing area is added between the modules. The global routes are determined with respect to the fixed capacities of these routing regions. In the following an integrated genetic approach to the optimisation of macro cell layouts is presented. Here a placement is encoded as a binary tree, and the exact positions of the modules are not fixed before global routing. Due to this there are no restrictions during the computation of the global routes and so the shortest paths are always chosen for the signal nets.

10.3.1 The genotype representation

A slicing floorplan is a rectangular floorplan which can be recursively partitioned into two parts by either a horizontal or vertical cut (Figure 10.4, left). The hierarchy of these cuts and thus the arrangement of the rooms in that floorplan can be defined by a binary slicing tree. As shown in Figure 10.5, for a particular placement the leaves of this tree represent the macro cells (blocks), and each inner node describes a partial placement (metablock) composed of the building blocks characterised by their successors.

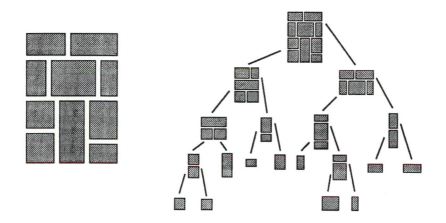

Figure 10.5   A placement (left) and its genotype representation as a binary slicing tree (right)

The binary slicing tree is an essential part of the genotype coding and the structure of this tree defines the relative placement of the blocks. The tree is constructed in a bottom up fashion by composing a metablock out of the two blocks characterised by the children of each inner node. Although the right block is always positioned upon the block defined by the left child, different orientations for both blocks are considered during the floorplanning process.

10.3.2 Floorplanning

When combining two blocks during the construction of the tree of an individual, their orientations are fixed. Both blocks are rotated to minimise the amount of wasted space inside the metablock, and to maximise the number of common nets on their channel borders. For flexible blocks, different shapes exist for the resulting metablock depending on the admissible implementations for both blocks. Fixing the implementations for the blocks contained inside the metablock would avoid this problem. Unfortunately, it is not clear at this time, which shapes are globally optimal. Only local decisions can be made, for example those implementations for the blocks may be chosen which minimise the wasted space inside this meta-block. Locally this is the best choice, but a different shape might be better to minimise the area of the whole layout. Due to this, all shapes for a metablock yielding from the combination of two flexible blocks after fixing their orientations are stored. Continuing this process down to the root of the tree does not produce an exponential growth for the number of stored shapes, because there are a lot of redundant implementations for the metablocks. Figure 10.6 presents the shape function for a meta-block that consists of two flexible blocks, one with two and the other with three different implementations, which yields six combinations. Because half of these possible shapes are covered by others, only three implementations for the metablock have to be stored.

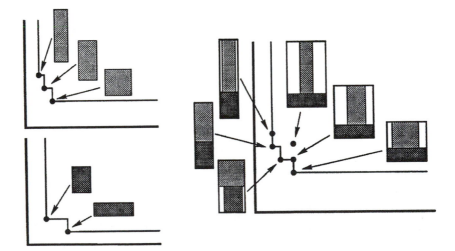

Figure 10.6   Combining the two flexible blocks yields three nonredundant implementations for the resulting metablock

After the construction of the last metablock, which is defined by the root, the tree represents a complete layout. Since there is a shape function for the root node, this single tree defines several layouts with different shapes. Out of these implementations, the layout with the minimal area can be determined. During a top-down traversal of the tree, in each inner node those implementations for both successors can be identified, and these add up to the optimal shape for the metablock represented by this node. After this traversal of the tree for all flexible blocks, those shapes which make up an optimal layout are chosen, i.e. the flexible blocks have been sized.

10.3.3 Integration of routing

Although global routes for the signal nets are determined later, during placement routing these can already be considered. When fixing the orientations of two blocks during the composition of a metablock, those orientations can be chosen which reduce the routing space inside that partial layout that is characterised by this metablock. For that purpose, the maximal channel width is estimated by inserting a track for each net, which has to connect terminals at both blocks, or – when combining metablocks – inside both metablocks. The number of tracks is then reduced by the number of nets, which have terminals on both sides of the channel. Thus the estimated channel width is minimal for those orientations that enable the direct connection of a maximal number of nets inside the channel. This channel width is added to the height of the metablock, and those orientations are chosen which yield a minimal area metablock. If the resulting metablock is flexible, all sixteen possible shape functions are computed, and the channel width is added to them (Figure 10.7). In this case those orientations are chosen that define the minimal area for the average of all implementations encoded in this shape function.

10.3.4 Computation of the global routes

After the construction of the tree and the floorplan, the routing graph as shown in Figure 10.8 (left) is constructed. As in the usual method of computing global routing (see Section 10.2.3), the routing space is now represented by a set of routing regions. Such a region may be a channel between two blocks, or a partition of a channel inside a metablock which combines two metablocks. In the latter case a region is a section in the channel which is bounded by two orthogonal adjacent channels, i.e. it describes the tract on which two cells have

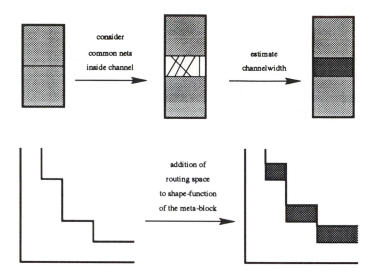

Figure 10.7   The insertion of routing space inside a metablock (top), and its addition to the shape function (bottom)

contact inside this channel. To compute a global route for a net, the shortest paths between its terminals in the routing graph are determined. In contrast to the usual computation of global routing as described in Section 10.2.3, the channels do not have any fixed capacities, but only an estimated width. The exact positions of the blocks are not fixed at this time, and thus only pseudo-optimal shortest paths are computed. After the computation of all global routes for each region the number of nets routed through it is known. For each net inside a region one track is added, which leads to the addition of an upper bound for the actual demand of routing space. The width of a channel is now set to the width of the widest of all regions that represent this channel. Because after floorplanning the exact terminal positions on all blocks are known, instead of adding a special track for each net, a better heuristic can be used to reduce the channel width.

When all channel widths are determined, a bottom up traversal of the slicing tree is done to add the channel width to the height of each metablock. Then the area of the routing space on the border of the layout is added to the layout area. After that, the shape for the root describes the total layout area, which defines the fitness of the corresponding individual.

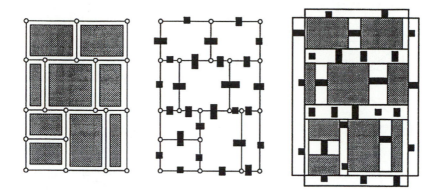

Figure 10.8 *A routing graph (left), the width for the regions after the computation of the global routes (middle) and the resulting channel widths (right)*

10.3.5 Hybrid creation of the initial population

Because of the large size of the search space it is profitable to start with a nonrandomly created initial population which already contains high quality building blocks. A special heuristic — iterated matching, introduced by Fritsch and Vornberger [7] — is used during the creation of the initial individuals to ensure that in each level of the tree those blocks or metablocks which share a maximal number of common nets are paired. In the first iteration, when the lowest level of the tree is computed, a complete graph is constructed: the vertices represent the blocks, and each edge is weighted with a value that describes the number of shared nets for the blocks characterised by its adjacent vertices. A matching in this graph is a set of edges such that no vertex is incident to more than one edge. Figure 10.9 presents an example with four blocks. In I – III the three matchings consisting of two edges are shown. The weight of a matching is the sum of the edge weights of the comprised edges. The maximum weight matching, II, characterises a set of block pairings with global maximal quality, i.e. with an overall maximal number of connected signal nets inside the partial layouts combined at this level.

In the second iteration, the next level of the tree is constructed by performing the same computation for a graph with vertices representing metablocks consisting of two blocks each. This process is iterated until the last two metablocks are joined at the root of the tree to build the complete layout.

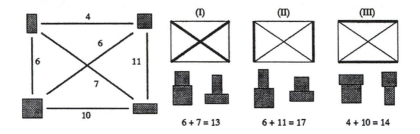

Figure 10.9   A matching graph with four blocks, three possible matchings and the maximum weight matching (II)

10.3.6 Crossover

Crossover is the most important search operator in genetic algorithms. It is a sexual recombination operator which constructs one or two individuals (offspring) out of the genetic information encoded in two parent individuals. The obvious way for the creation of offspring out of two tree-structured individuals is to combine some disjunct subtrees of both parents to a tree for the offspring.

Figure 10.10 presents the application of the crossover operator. Out of two parent individuals one offspring is produced. A set of disjunct subtrees in both parents is chosen randomly and builds a pool of building blocks out of which a new individual is composed. If the leaves of these subtrees do not represent all modules of the circuit, the missing blocks are inserted into this pool. During the composition of the upper levels of the offspring tree, iterated matching can be applied again. For the newly created metablocks, the orientations of the combined blocks are fixed to yield minimal area shapes as described in Section 10.3.3.

10.3.7 Mutation

In simple genetic algorithms, mutation is the less used but not the less important operator in comparison to crossover. Each gene in an offspring which has been created by crossover can be mutated with respect to the mutation probability (or mutation rate) which is usually proportional to the size of the problem (length of the chromosome). The mutation operator produces genetic information, which might be new to the population, or reintroduces information that has been eliminated by selection, but might be helpful during the current state of the search process.

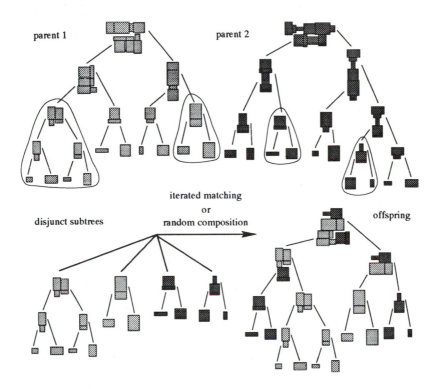

Figure 10.10 The crossover operater chooses disjunct subtrees in both parents and combines them to form an offspring tree

There are three mutation operators which are applied with different frequencies. They change the structure of the slicing tree by exchanging leaves or subtrees, moving subtrees to other positions in the tree, or changing the orientation of blocks or metablocks. Figure 10.11 (top) shows the effect of an operator which exchanges two parts of the tree. Part A represents a single block, and subtree B contains two blocks. On the phenotype level, this corresponds to exchanging the cell defined by block A with the placement for the set of cells characterised by subtree B. The second mutation operator (Figure 10.11, bottom) picks up a part of the tree (block A) and inserts it at a different position (x). This corresponds to cutting a cell or a partial layout out of the complete layout and moving it to a different place. The last mutation operator changes the orientation of a block or a metablock inside a tree. As mentioned in Section 10.3.3, the orientations have been fixed during the construction of each new metablock due to local decisions, i.e. those orientations have been chosen which add up to a minimal area shape for this metablock. However, globally a different orientation for a block or metablock might be better.

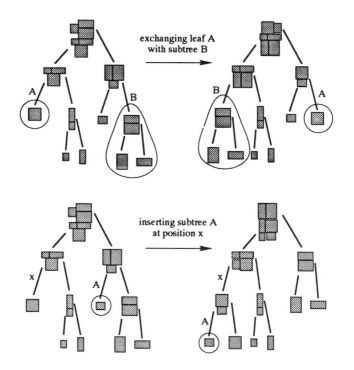

Figure 10.11 The two basic mutations: exchanging (top) or moving (bottom) subtrees

10.3.8 Selection

For the mating of individuals during crossover, truncation selection is used, i.e. only the better individuals are selected for recombination. For mutation any of the individuals in the population can be chosen with equal probability. A steady state type of GA is used: thus an individual may survive for longer than only one generation. At the end of each generation individuals are replaced if the quality of the offspring is better.

10.4 Results

The algorithm has been tested on real-life circuits chosen from a layout benchmark suite which was released for design workshops in the early 90s and is often referenced in the literature as the MCNC-benchmarks. The benchmarks were originally maintained by MCNC (North Carolina's Microelectronics, Computing and Networking Center), but are now located at the CAD Benchmarking Laboratory (CBL) at North Carolina State University [1]. These benchmarks are problems from the field of full custom macro cell layout; the characteristics of the circuits are shown in Table 10.1. Although circuits *xerox*, *ami*33 and *ami*49 are original problems, *ami*33_3, *ami*49_3, *ami*33_5 and *ami*49_5, are adaptations of these circuits with all cells having three and five different shapes, respectively.

Table 10.1   The benchmark circuits

	xerox	ami33	ami33–3	ami33.5	ami49	ami49_3	ami49_5
#cells	10	33	33	33	49	49	49
#shapes per cell	1	1	3	5	1	3	5
#nets	203	123	123	123	408	408	408
#terminals	698	452	452	452	958	958	958
# I/O terminals	2	42	42	42	22	22	22
cell area [mm²]	19·4	1·16	1·16	1·16	35·1	35·1	35.1

All results presented in this Section are computed on a Parsytec GC/PP parallel computer with Motorola MPC 601 PowerPC processors. The parallel version of the GA executes the sequential GA on a number of islands, with individuals migrating between these islands every three generations. Each island holds a subpopulation of ten individuals.

Figure 10.12 shows a layout for circuit *ami*49, a problem with only fixed-size cells. The layout does not include the 22 pads and the interconnection wirings for the nets which are connected to them. As can be gathered from Table 10.1, nearly all of the signal nets are two terminal nets. The total layout area is 58·9mm²; about 40% of this area is not occupied by the cells and thus represents routing space or wasted area.

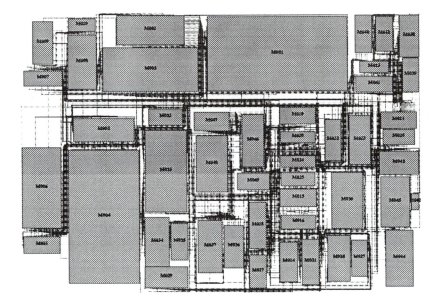

Figure 10.12 Layout for ami49 generated by the GA

Table 10.2 shows statistical data for a set of 1000 randomly generated initial layouts for circuits with fixed macro cells. It shows the benefit of iterated matching during the creation of the initial individuals. It can be seen that the standard deviation in those cases where the individuals are created through the use of this heuristic is quite small compared to this value for randomly generated trees. This is because iterated matching is a deterministic process that always pairs the same blocks in each individual. Variation is only caused by routing because here different orientations are used for the blocks during their composition.

Table 10.2   *The statistics for 1000 generated individuals showing the use of the*
iterated matching heuristic

area	xerox		ami33		ami49	
[mm²]	random	match	random	match	random	match
best	33.9	49.8	5.78	6.45	85.0	85.7
worst	76.3	52.5	16.75	9.41	226.1	115.4
avg	45.5	51.0	10.16	7.51	132.2	98.0
σ	6.4	0.7	1.83	0.68	23.9	8.2

Figure 10.13 presents the performance of the genetic algorithm for circuit *ami* 49; the fitness of the best individual averaged over ten runs is shown. The upper curve describes the progress of the layout area for runs of the algorithm starting with populations of totally randomly generated individuals and does not use iterated matching during crossover. The lower curve describes the performance of runs which have started with a set of individuals that have been generated by application of iterated matching, and iterated matching is used during crossover, too. For this circuit the convergence of the GA is clearly better when using the iterated matching heuristic. In comparison to

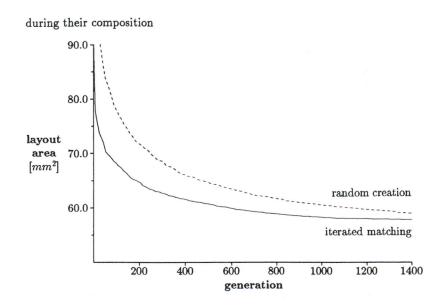

*Figure 10.13 Performance with respect to the iterated matching heuristic for the
creation of the initial individuals and during crossover (ami49, avg. of
ten runs each, 16 islands)*

this, for circuit *ami33* (Figure 10.14) iterated matching is only beneficial in the early stage of optimisation. Here, after 300 generations, the version of the GA that randomly pairs blocks during recombination clearly outperforms the other. Thus it would be advantageous to adapt the use of iterated matching during the optimisation process, depending on the number of blocks, the number of generations, and the success of the recombination operator.

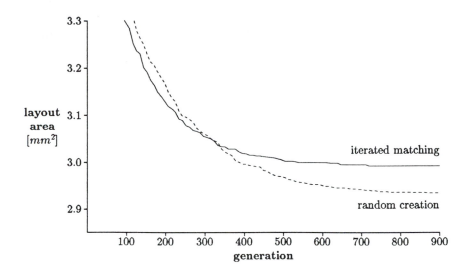

Figure 10.14   The performance of the algorithm for circuit ami33 (avg. of ten runs, 16 islands)

Table 10.3   The maximal and average number of shapes for the root block and all metablocks of 1000 randomly generated individuals

#shapes	ami33_3	ami33_5	ami49_3	ami49_5
max	39	75	58	92
root	18·6	51·6	40·7	63·1
avg	6·9	13·8	8·9	14·6

As stated in Section 10.2.2, for circuits with flexible cells it is useful to store all nonredundant layout alternatives for the metablocks in shape functions. Table 10.3 presents the number of shapes for the root block and for all metablocks of 1000 randomly generated trees. It can be seen from these values that even for the circuit *ami49_5*, where each of the 49 blocks can take five different shapes, such an approach is practicable since the average number of root shapes is only 63. Figure 10.15 shows the benefit of storing shape functions for the metablocks for circuit *ami33_3* with 33 flexible cells, each having three different implementations. Considering Table 10.3 one can assume that for this circuit each individual represents a layout with 18·6 shapes on average and that the average number of shapes in an inner node might be about seven. In comparison to the performance of the full version of the GA (lower curve), the performance of runs without shape functions for the metablocks is shown. In this case (upper curve), when combining two flexible blocks, only the locally optimal shape for a metablock is stored, i.e. the shape with minimal area (minimal waste inside this metablock). Obviously, the version which stores all alternatives for the metablocks clearly outperforms the other one.

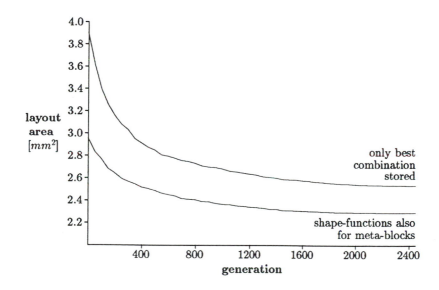

Figure 10.15   The benefit of shape functions for metablocks, average of ten runs with eight islands (ami33_3)

10.5 Conclusions

In this Chapter, a genetic algorithm for a real-world combinatorial optimisation problem – the design of VLSI macro cell layouts – has been presented. The main feature of this approach is the total integration of global routing into the placement process. During floorplanning estimated channel widths are added to the shapes of the partial layouts; after sizing, the positions of the cells on the layout surface are still flexible. Thus, computation of the global routes can be done without any restrictions, and an individual defines a completely routable placement.

To enable the GA to deal with this complex optimisation problem, many details have been incorporated, the essential differences of this approach in contrast to simple GAs being:

- A problem-specific genotype representation as a binary tree.
- Nonstandard genetic operators which directly work on the tree structure and only produce admissible individuals.
- Implicit optimisation due to the use of problem-specific knowledge during the construction of the individuals.
- A special heuristic to create high-quality individuals for the initial population and during recombination.

When designing a genetic algorithm for a complex optimisation problem, most effort has to be put into finding a proper coding for the solutions to the problem. There exist several examples in the literature for possible representations one can choose. Mitchell summarised this in her recent book ([11], p. 155): '... *when one wants to apply the GA to a particular problem, one faces a huge number of choices about how to proceed, with little theoretical guidance on how to make them.*' In common GAs the chromosome, which represents the coding, is a string of elementary datatypes. Often binary strings are chosen, because these were originally motivated by Holland's minimal alphabet theory to enhance the implicit parallelism [9]. Most of the theoretical work on GAs is based on binary coding. For many continuous parameter optimisation problems a string of floats, like that used in the *Evolution strategies* [16], is the better choice. But for combinatorial problems such as macro cell layout optimisation, a more complex coding has to be used. Since the search space is discrete, care has to be taken to generate only admissible individuals during the optimisation process. For that purpose, problem-specific operators have to be designed which make use of problem specific knowledge. During offspring creation, this knowledge can be used for implicit optimisation.

One of the main properties of problems like the one presented in this Chapter is that even partial solutions or parts of the chromosome can be evaluated. For example, during the composition of a meta-block, all possible orientations for both blocks are checked, and the orientations yielding the highest quality metablock of all are chosen. Here a local decision is made depending on the fitness of a partial solution. In contrast to this is the composition of two flexible blocks to a metablock. At that time, the globally optimal shape for the metablock cannot be determined. Storing all nonredundant implementations for the metablock in a shape function is an example of another way to include problem specific knowledge to support the GA to create high quality individuals during the optimisation process. However, when using hybrid techniques, there is always the risk of only incompletely sampling the solution space, as was shown for the use of iterated matching. Although it was very beneficial for one problem, it was disadvantageous for another. Additionally, the profit of incorporating heuristics might depend on the stage of the optimisation process. To handle this, adapting the use of the heuristic during the search is necessary.

Enhancing the performance of the search can also be done by restricting the search space, as is done during routing. Before including deterministic computation of the global routes based on shortest paths, research was done on optimising the routing using the GA together with the placement [14]. The benefit of that approach was the fact that, e.g. after recombination, the routes inside the inherited partial layouts do not need to be recomputed, only the routing for nets connecting terminals in both parts has to be done. Optimisation of the global routes using the GA is certainly possible, provided that special operators for modification of the routing are added. Because of the enormous growth of the search space due to the huge number of possible routes for each net, this approach is impracticable if one expects the GA to produce a good solution within a realistic time bound. Of course, computing the shortest paths in the current implementation takes more than half of the total runtime of the GA, but finally it saves a lot of time, since it drastically reduces the number of generations needed before converging on an optimal solution.

10.6 Acknowledgments

This work has been supported by the German Federal Ministry for Education, Science, Research and Technology (BMBF) as part of the project 'HYBRID—application of parallel genetic algorithms in combinatorial optimisation' under grant 01 IB 405 E3. The author thanks the Paderborn Center for Parallel Computing (PC²) for the opportunity to use the parallel machines located there. Special thanks to Oliver Vornberger, Frank Lohmeyer and Frank M. Thiesing for many helpful suggestions.

10.7 References

1 CBL, CAD Benchmarking Laboratory, North Carolina State University, http://www.cbl.ncsu.edu/www/CBL_Home.html.
2 Chan, H,. Mazumder, P., and Shahookar, K.: 'Macro-cell and module placement by genetic adaptive search with bitmap-represented chromosome', *Integr. VLSI J.* **12,** pp. 49–77, 1991
3 Cohoon, J. P., and Paris, W. D.: 'Genetic placement', Proc. of IEEE int. conf. on *CAD*, pp. 422–425, 1986
4 Cohoon, J. P., Hegde, S. U., Martin, W. N., and Richards, D. S.: 'Distributed genetic algorithms for the floorplan design problem', *IEEE Trans.*, **CAD 10**, (4), pp. 483–492, 1991
5 Esbensen, H.: 'a genetic algorithm for macro cell placement', Proc. of the European *Design Automation* Conference, pp. 52–57, 1992
6 Esbensen, H., and Mazumder, P.: 'SAGA: A unification of the genetic algorithm with simulated annealing and its application to macro-cell placement'. Proc. of the 7th int. conf. on *VLSI design*, pp. 211–214, 1994
7 Fritsch, A., and Vornberger, O.: 'Cutting stock by iterated matching', in *Operations research proceedings, selected papers of the int. conf. on OR 94,* Derigs, U., Bachem, A., Drexl, A. (eds.) (Springer Verlag, 1995), pp. 92–97
8 Goldberg, D. E.: *Genetic algorithms in search optimisation & machine learning* (Addison-Wesley, 1989)
9 Holland, J. H.: *Adaptation in natural and artificial systems* (MIT Press, 1992)
10 Lengauer, T.: *Combinatorial algorithms for integrated circuit layout* (John Wiley Sons, 1990)
11 Mitchell, M.: *An introduction to genetic algorithms* (MIT-Press, 1996)
12 Otten, R.: 'Efficient floorplan optimisation', Proc. of int. conf. on *Computer Design*, pp. 499–502, 1983
13 Sait, S. M., and Youssef, H.: *VLSI physical design automation: theory and practice* (McGraw-Hill, 1995)
14 Schnecke, V., and Vornberger, O.: 'Genetic design of VLSI-layouts'. Proc. first IEE/IEEE int. conf. on *GAs in engineering systems: innovations*

and applications, GALESIA '95, IEE Conference Publication No. 414, pp. 430–435, 1995

15 Schnecke, V., and Vornberger, O.: 'A genetic algorithm for VLSI physical design automation'. Proc. 2nd. int. conf. on *Adaptive computing in engineering design and control*, ACEDC '96, University of Plymouth, Parmee, I. C. (ed.), pp. 53–58, 1996

16 Schwefel, H.-P.: *Evolution and optimum seeking* (John Wiley & Sons, New York, 1995)

17 Sechen, C., and Sangiovanni-Vincentelli, A.: The timberwolf placement and routing package', *IEEE J. Solid-State Circuits*, **SC-20**, pp. 510–522, 1985

18 Shahookar, K., and Mazumder, P.: 'VLSI cell placement techniques', *ACM Comput. Surv.*, **23**, (2), pp. 143–220, 1991

19 Sherwani, N.: *Algorithms for VLSI physical design automation* (Kluwer Academic Publishers, 1993)

20 Wong, D. F., Leong, H. W., and Liu, C. L.: *Simulated annealing for VLSI design* (Kluwer Academic Publishers, 1988)

Index